MANAGING PROJECTS IN TELECOMMUNICATION SERVICES

MANAGING PROJECTS IN TELECOMMUNICATION SERVICES

Mostafa Hashem Sherif

IEEE Communications Society, *Sponsor*

IEEE PRESS

A JOHN WILEY & SONS, INC., PUBLICATION

Published by John Wiley & Sons, Inc., Hoboken, New Jersey.
Published simultaneously in Canada.

For general information on our other products and services or for technical support, please contact our Customer Care Department within the United States at (800) 762-2974, outside the United States at (317) 572-3993 or fax (317) 572-4002.

Wiley also publishes its books in a variety of electronic formats. Some content that appears in print may not be available in electronic format. For information about Wiley products, visit our web site at www.wiley.com.

Library of Congress Cataloging-in-Publication Data is available.

ISBN-13 978-0-471-71343-2
ISBN-10 0-471-71343-0

Printed in the United States of America.

10 9 8 7 6 5 4 3 2 1

CONTENTS

FOREWORD

If ever there was a "tip of the iceberg" in industry, telecommunications is it. Probably just a few people in the world wouldn't know what telecommunications are, most from direct experience, some from hearsay. However there are even less people who appreciate what behind the scene in telecommunications is like; what is the magic that lets anyone pick up a phone almost wherever on the Earth and with a few pushes on the phone's keys, connect with people on the other side of the globe or just few supermarket shelves away, as so often is the case.

Now, if you ask people about this magic, probably 100% will tell you that, although they do not understand the what and why, surely technology is what makes this telecommunications clock tick. None is likely to think about the clock master, the people that run the show. And what a show it is.

We have in Italy (you can do the multiplication for the worldwide figures) 40 million pairs of fixed lines to reach 26 million users. Add 56 million clients using cell phones connected via more than 20,000 base stations. The length of copper and fiber lines is measured in millions of kilometers, and you've got to know where each single meter of cable lies. You have to know about the hundreds of thousands of pieces of equipment, keeping track of thousands of maintenance vans; take into account the skill of every one of the tens of thousands of people making the telecommunications magic invisible to the users. And once you have taken all of this into account, you have not even started yet.

Some of my friends are from the computer world and they use to brag about how good they are when you consider innovation and speed of innovation. Their COMDEX meeting is a twice-a-year event; by comparison the ITU Telecom Fair (the planetary equivalent of COMDEX in telecommunications) takes place once every four years. They tell me,

"Look at computer shops windows. Every other month there is something new on the shelves." In telecommunications everything seems to evolve at a snail's pace.

Actually this is all true and completely wrong. The view we take at the computer world is a micro view: We look at the single PC and at the new models coming out. If we were to look at all the PCs disseminated in offices and homes, we would see an evolution that is much slower. It takes probably four years to find a significantly changed environment. Same goes for telecommunications. We usually look at them in a holistic way and as such evolution is perceived in years. Were we to look at micro level, at single equipment, we would discover that not a single week goes by without a significant update in the network.

There are literally hundreds of services and a beehive of activities every single day in deploying, testing, activating, maintaining these services. Now we are getting closer to the foundation of the "magic." Design of new services and products is a never-ending activity in telecommunications. And in the last 10 years, it has become even more so, with no sign of relenting in the next decade; rather the opposite.

It has been said that telecommunications are neutral, their goal is to provide platforms that anyone and any business can use. Indeed that is the case. But more and more, serving the business and the variety of people's needs requires a much broader view. Designing a new service has to take into account what people want, what kind of terminals they have and are likely to have in the next few years, what kind of infrastructure is and will be available here and on the other side of the world. Such a task cannot be handled by one engineer. It requires a variety of skills, the blending of competence coming from different people that have to cluster into a team.

It is the success of this team that leads to the success of the industry in the market place. And this is what this book is about. Tearing away the veil of magic to show how we can make sure that the magic will continue on and on.

Of course, it is not about magic. It is about "biz." The U.S. voice market in 2005 has exceeded $100 billion. That voice biz is expected to decline in the coming years under competitive pressure and shrunk by the Internet VoIP plus unlicensed providers deploying myriads of access networks. The challenge is to keep the overall telecom biz at the current levels, in spite of that decrease (optimists would rather say to increase the present telecom biz). This can only be done by offering new services, by entering into vertical markets, by supporting other value chains in increasing their productivity and thus diverting part of the decreased cost to telecoms' pockets.

All of this requires a deep understanding of processes, of the tasks involved in a service life cycle, from inception to demise. It will require project management at its best . . . read on.

ROBERTO SARACCO
Trends and Scientific Communications Director
Telecom Italia

PREFACE

Bringing into existence an idea or a conceptual design is typically a collective endeavor of a group of people with the necessary expertise and dexterity. The efficacy of such an ad hoc association in achieving its mission can be fostered by conscious steps to optimize the organization and partition of work, to harmonize organizational objectives with individual opinions, to smoothen internal conflicts, and to navigate external pressures. The logical consequence is that a balanced mixture of "hard" and "soft" knowledge gained from disciplines ranging from engineering and statistics to industrial psychology and anthropology can improve the conduct of projects and enhance their chances of success.

The worldwide wave of deregulation has exposed hitherto vertically integrated structures to competitive forces. Previously impregnable positions have become exposed as the workflow was restructured to delocalize activities and to outsource peripheral functions. While production systems are becoming more complex and interdependent and their potential failures more compelling, the organizational interfaces are being revised, updated, or re-engineered. Not surprisingly, project management was propelled to the fore as a disciplined approach to face change, undertake continuous improvement, navigate cross currents, and reduce susceptibility to risks. The last decade has witnessed a quantitative jump in related publications, conferences, and academic programs, along with a worldwide surge in the number of professionals certified by the Project Management Institute (PMI); some companies even reorganized around project teams.

Even though in many countries a large majority of the workforce is now engaged in service activities and, particularly in the Western world, the contribution of the service sector to the Gross National Product (GNP) is more than twice that of manufacturing (around 68% versus 30% respectively), projects in industrial design and production have captured most of the interest. The literature on services in general, and telecommunication services in particular, is rather sparse, with discussions mostly focused on the competitive advantages of advanced telecommunications and a global information infrastructure. In fact, I was able to locate only one title in English devoted exclusively to the subject: *Telecommunications Project Management* by James B. Pruitt published in 1987 and now out of date and out of print, although Ellen Ward (1998) provided an excellent description of the service development process and Celia L. Desmond (2004) gave a reference guide for business managers to the telecommunication environment. Stéphane Calé (2005) provides some up-to-date discussions on the management of risks, quality, and faults in modern networks, but his book—in French—is restricted to enterprise networks.

An unfortunate consequence of this lacuna has been the lack of awareness of the specifics that can make or break projects in telecommunication services. To many decision-makers, it is sufficient to deploy advance equipment, string them with fiber optics, and run popular applications. From that perspective, the service of selling hotdogs or delivering milk is indistinguishable from that of offering a reliable, affordable, and ubiquitous telecommunication infrastructure.

This book is an attempt to fill this gap; its ambition is to provide an integrated methodology to help managers of projects in telecommunication services to make informed decisions. It is based on extensive real-world experience with diverse telecommunication projects to provide practical insight on the issues that face project managers. It draws on a wide range of disciplines from telecommunications to organization management, as well as from motivation to quality control and software reliability engineering. This is not, however, a compendium of recipes, because in a rapidly changing world, it is not possible to figure out all situations and chart out step-by-step solutions to the various combination of problems that may arise.

The book is organized in three main parts as follows: Chapters 1–3, 4–10, and 11–12. The purpose of the first part is to position telecommunication services taking into account the technology life cycle, the type of innovation, and the project organization. Chapter 1 gives what separates projects in service development from those for equipment development, a distinction that escapes many. Chapter 2 extends the literature on innovation and the technology life cycle to the area of telecommunication services, with a special emphasis on standards as essential building blocks for end-to-end service transparency. Chapter 3 describes the need to tailor the project organizational arrangement to the type of innovation to ensure that the right information reaches to the right people at the right time.

Chapters 4–10 cover the areas that the PMI had standardized in its publication *A Guide to the Project Management Body of Knowledge* (PMBOK® Guide). The subjects are discussed in the following sequence: scope, schedule and cost, information and communication, human resources, quality, vendor management (procurement), and risk. Rather than repeating material readily available in the hundreds of books on general project management, the focus will be on the issues specific to telecommunication services and supplements to the classical lore from the literature on the management of technical innovation. Inevitably, the treatment is based on my own personal experience (because the corpus of the subject is still fluid); the hope, however, is that the material presented will be useful to a large number of telecommunication project managers. Chapter 11 is an integration of all these concepts for the planning and delivery of a project. Chapter 12 summarizes the main ideas of the book with a look toward trends.

I have tried to write the chapters in such a way that the reader can read them independently. This is why some material or discussions are repeated in more than one chapter. Hopefully, this will reinforce some of the key points on tying the innovation to the technology life cycle, the effect of culture on project management, or risk assessment. If not, I would then ask the reader's indulgence and advise on how to make the repetition less tedious.

The first time I realized the need for such a book was in 1996–1997 while working in Turkey as part of a team from what was then called AT&T Value-Added Services. My collaboration and long discussions with Steve Pollack and Alex Zwahlen influenced the notes that formed an early precursor of Chapter 11. Chapter 8 incorporates quality practices developed and tested on real projects with David Hoeflin and Michael Recchia.

Throughout the years, I was fortunate to meet and have discussions with numerous distinguished colleagues from many companies and various countries, in academia, industry, and standard organizations. I attempted to distill their insights, for which I remain deeply grateful, into the various chapters. Their list is too long to mention by name, lest I omit some by mistake. Therefore, I will restrict the acknowledgments to those who directly commented on the various drafts of the book. They are: Professor Audrey Curtis from Stevens Institute of Technology; and Rod Castillo, Robert J. Ferro, Thomas K. Helstern, Dr. Clement McCalla, and Fahad Najam from AT&T. Michael Recchia, my previous supervisor, gave me useful suggestions on the treatment of risks. Cathy Savolaine, a retired department head at AT&T Bell Laboratories and AT&T Laboratories, read the whole manuscript twice and recommended the addition of what is now Chapter 12. The title came from Fred Burg, another retired colleague from AT&T, over coffee and bagels after a weekend run.

At one time, AT&T had an excellent research library. I was lucky to collect most of the necessary data before AT&T management unfortunately decided that they no longer needed a first-class reseach library. Jane Bogdan, who was the librarian at that time, helped in locating hard-to-find references; so did Hsiao-Chuan (Cathy) Wu from AT&T Information Research Center.

Catherine Faduska, Senior Acquisitions Editor, guided the book proposal through my publisher, Wiley-IEEE Press.

MOSTAFA HASHEM SHERIF

Tinton Falls, New Jersey
July 2006

1

PROJECTS IN TELECOMMUNICATION SERVICES

INTRODUCTION

The telecommunications industry spans many different activities that fall into two main categories: building equipment and using that equipment to connect people and machines. Both equipment manufacturers and service providers are regulated in one way or another, although the degree of regulation varies with the dominant political ideology as well as the markets. While at a certain level of abstraction, all projects can be treated with the same approach, generic techniques need to be supplemented with scrupulous attention to details specific to the industry; one approach cannot fit all possible situations, and customization or original development may be needed for specific situations. The purpose of this chapter is to increase the reader's understanding of the nature of projects in telecommunication services and highlight what distinguishes them from those in equipment design and development. We start by clarifying a few terms from the project literature and then provide several examples to illustrate the common characteristics of service projects. We end the chapter by contrasting projects in telecommunication services with those that are related to the development of equipment.

Project Management Versus Product Management

Projects are temporary endeavors undertaken to create a unique product or service. *Project management* is the application of knowledge, skills, techniques, and tools to align the resources and skills needed to achieve the objectives of the project within specific constraints of cost, time, and quality. It portrays a disciplined approach to integrate various data elements describing the project and draw a coherent picture of its status to guide decisions. *Product management,* in contrast, is related to all aspects of a product line including life-cycle management of existing products as well as the development of new products to achieve a competitive advantage [Gorchels, 2003].

Historically, the discipline of project management was applied first in construction, then in large government projects such as defense, as well as in the chemical and pharmaceutical industries. Its application expanded later to other fields such as software development and telecommunications. In the case of software development, it was observed that 15–25% of all projects failed to complete and that there was great dissatisfaction with the quality, cost, or timeliness of those that did complete. This has spurred the adoption of project management in the conduct of software projects as a way to bring runaway processes under control. Likewise, the increased interest in formal project management techniques in telecommunication services can be attributed to several factors. Changes in the regulatory regime have imposed the unbundling of many telecommunication services into their individual constituents, while several new technologies have become available to service providers. This combination of regulatory and technological changes has led to an increase in the number of potential suppliers as well as candidate solutions. The multiplicity of choices at each level of the service hierarchy (infrastructure, network, application, content) has made the interactions among vendors, sponsors, and customers extremely complex, especially when some of the operational tasks are outsourced. Finally, competitive pressures are forcing service companies to deliver their products faster, with higher quality and with lower cost. Conducting successful projects in such a dynamic and risky environment requires the discipline that formal project management fosters.

The current architecture of telecommunication services is shown in Figure 1.1. An infrastructure provider is responsible for making transmission bandwidth (fiber cables, undersea cables, satellites, etc.) available. A network provider builds, operates, and maintains the network elements and infrastructure. The service provider buys network services from a network provider and then resells them to end-users, other service providers, and content providers. The service provider can be an Internet service provider (ISP), a provider of disaster recovery or a storage area network (SAN) provider, a call center operator, a web host, and so on. The content provider is responsible for content creation and can consolidate catalogs (e.g., directory services), store voice messages, provide answering services (call centers), or provide digital certificates. Finally, among the functions of the content manager would be managing customer relationship, packaging contents from several content providers, facilitating electronic payments, acting an exchange or a market place for electronic commerce, storing content, and so on. Clearly, many independent entities have to cooperate to integrate their particular subcomponent in an end-to-end service offer. Furthermore, the planning and development of infrastructure projects can last several years and could involve up to several thousand persons from many suppliers. Project management techniques are needed to prevent the fragmentation that may plague large-scale engineering projects and to ensure that [Bergren et al., 2001]:

- The activities of the various parties remain coordinated without unnecessary rigidity or bureaucracy.
- The project activities remain relevant through controlled of the changes to the technical and quality requirements to track the environment or the customer needs.

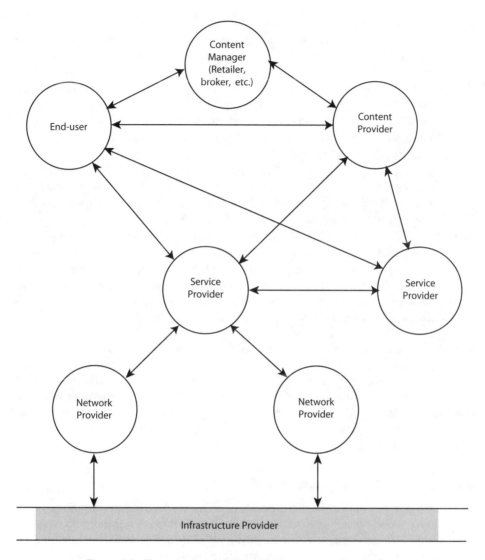

Figure 1.1   The current architecture of telecommunication services.

- Experience gained is recorded and transmitted to improve organizational learning, even though techniques of lean management make such learning very difficult.

Virtual Network Operators

Service providers can be viewed as virtual network operators (VNO) to underline the fact that they have no physical assets and that they buy the connectivity from specialized suppliers. In turn, they concentrate on the management of customer relations as well as supplier management. Let us consider some examples.

- AT&T's consumer long distance could fit that business model because it does not own physical assets. Its services run over its parent's national and transoceanic networks as well as the local access networks of the various telephone companies within and outside the United States. Its principal activity is to manage 50 million customer relationships through its customer care systems [Martin, 2005, p. 142].

- Call-back operators are another example of service providers. When the cost of a call from Country A to Country B is higher than the cost of a call in the opposite direction, callers from the first country can reduce their bill by asking their correspondent to initiate the call. This was turned into a business by having an agency in Country B offering the service for a fee to those that have an account with it [Wheatley, 1999, pp. 411–412].

- In the data area, @Home was an ISP created by cable companies to offer broadband access to the Internet over local cables in exchange of a share of its revenues. The content it distributed was stored in 25 regional data centers connected through a backbone network provided by the long-distance companies.

- Vonage and AT&T's CallVantage^{SM} service offers of telephony over IP networks are services that are offered over the broadband connections of cable companies. In addition, the network provider that offers IP connectivity can be different than the service provider. This allows services such as *virtual phone numbers* whereby incoming calls can be routed over IP connections anywhere in the world. Thus, people can make an international call with the price of a local call.

Virtual companies may even restrict themselves to customer management and marketing, leaving to their suppliers all other technical aspects of the service. The quality of the service in terms of availability or billing accuracy becomes highly dependent on the network operator. New services such as electronic commerce or geolocalization must fit within an environment that already exists. However, such a business model is vulnerable to the suppliers' cooperation and decision not to offer competing services using the physical and network management infrastructure that they are leasing.

Contribution of Project Management

As discussed above, project management is becoming essential for a more efficient service delivery process that minimizes the risks of cost overruns or schedule slippages and increases the chances of success. In particular, project management aids in assessing the value of the project implementation and providing proactive guidance on the conduct of the implementation with objective metrics to answer the following questions [Thorp, 1998, pp. 51–52]:

1. Are we doing the right things?
2. Are we doing them the right way?
3. Are we getting them done well?
4. Are we getting the benefits?

In other words, project management supplies tools to do the following:

- Circumscribe the scope of the project and any changes to that scope.
- Define and maintain communication links across organizational and occupational boundaries.

- Anticipate risks and uncertainties.
- Measure progress and the quality of the work delivered.
- Acquire knowledge through experience and share it among the project team.
- Ensure accountability.

THE TWO FACETS OF TELECOMMUNICATION SERVICES

A common property of service offers is that they have two facets, depending on whether they are seen from the end-user viewpoint or with an eye on internal operations.

The external view of a telecommunications service offer is that it consists of services that are available to subscribers to link them with each other. The nature and the characteristics of the services vary according to the customer segment, whether it is for a mission-critical business application, for an enterprise network, or for the mass market. The interval view of telecommunication services concerns the capabilities, processes, and functions that allow the organization to deliver that service. This includes the networking technologies in addition to operations support systems, methods and procedures, applications, and content distribution. Figure 1.2 shows the internal view of the components of telecommunication services. Thus, the design of a telecommunication service includes the networking technologies, the operation support systems (OSS), and the management of procurement, testing, installation, operation, maintenance, and billing of telecommunication services. Accordingly, the scope a project in telecommunication services is to bring these components together from concept to life-cycle management, even though these components are not synchronized in their life. In one area, a technology may still be under development while other technologies in another area may be mature or near retirement.

The *networking technology* component relates to (a) the physical infrastructure involved in the end-to-end traffic delivery such as cables and transmission lines, (b) the network elements such as switches and routers, (c) software-enabled capabilities such as messaging and call forwarding, and (d) networked applications such as web hosting or

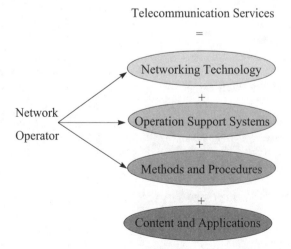

Figure 1.2 Elements of a telecommunication service.

storage networks. The equipment includes multiplexers, cross-connect, routing and switching equipment, power systems, and security systems such as intrusion detection systems. In some cases, such as managed services, customer premise equipment will be included. The networks may vary in complexity, size, the technologies used and their interaction with each other, their topology, and so on. Innovations at the product level provide media excitement and stimulate enthusiasm for the diffusion of technology. However, from a service viewpoint, what counts is the ability to deliver and maintain the quality of the end-to-end service.

The *operation support systems* component relates to the various network element management systems as well as systems used for provisioning, accounting, security, billing, and so on. This component is essential for development, deployment and maintenance of high-quality network-based services using shared facilities, such as for public networks. It should be noted that these systems are less important in the case of private and enterprise networks.

The *methods and procedures* are routines to streamline the many tasks needed for (a) the installation of equipment, (b) the engineering of the network, (c) managing the maintenance and repair operations, and (d) the customer support functions. It is true that the various service providers share more or less the same technology and have similar support systems; furthermore, standardization is essential to ensure end-to-end connectivity and to reduce the complexity of service management. Nevertheless, the distinctive advantage of any service provider resides in is capacity to attract and retain customers and to support growth to reach profitability. In other words, the more the technology is standardized, the more important is the service delivery process in the market success of telecommunication services providers [Ward, 1998].

Finally, the *contents and applications* refer to content creation (e.g., customer relations management, disaster recovery, electronic data interchange, etc.) or the packaging or consolidation of contents from several sources. This content can be news, movies, voice mail, web hosting, weather reports or stock price, voice messaging, taxi services, catalogs, certificate management for electronic commerce, and so on. This area is typically outside the focus of network operators or service providers, even though the availability of content is essential for the success of data telecommunication services. For example, to stimulate the growth of some telecommunication services, such as i-mode of NTT DoCoMo or the older Minitel of France Telecom, the network operator used its direct relationship with the end-user to act as an intermediary for content providers—that is, as a distributor and to collect payments. In that role, the operators added the amount due to the content provider to their monthly bills. With the growth of mass markets for the exchange of digital pictures and music files through peer-to-peer sharing as well as the widespread usage of mobile camera-phones, end-users can now act as content providers, like they were in traditional telephony.

The issues related to the management of content generation, particularly when it touches intellectual property rights, will not be considered in this book.

CATEGORIES OF PROJECTS IN TELECOMMUNICATION SERVICES

Telecommunication services vary according to several factors. These include the nature of the network used (public, private, or virtual private), the target market (consumer, business, government, military, emergency services, etc.), the nature of the installation (permanent or temporary), or the type of services (voice, entertainment, mission critical applications, etc.). To illustrate the characteristics of telecommunication projects in the area of services, let us consider the following examples

1. Upgrading the capabilities to an existing public network
2. Establishment of specialized business networks
3. Installation and dismantling of temporary networks

Upgrades of Public Networks

This category of projects relates to the replacement of obsolete technology, the deployment of a new service, and the enrichment of an existing service with new features, applications, or capabilities. The origin of these improvements may be new regulations, capacity growth to meet customer's demands, or the emergence of new technologies. Because these projects affect the general public, the challenge that faces network operators is to minimize disruptions to the existing services—that is, preserving a 24-hour-a-day, 7-day-a-week operation even when an obsolete technology is being replaced.

- Digital telephony (switching and transmission) replaced analog equipment transparently. Another example is the replacement of the processor of all the 135 4ESS switches in the AT&T voice network during the 1990s without downtime or service interruption [Golinski and Rutkowski, 1997].
- Examples for capacity expansion include the addition of new area codes or the changing of the numbering schemes to 10 digits in many countries, the migration of existing traffic to different transmission facilities such as a new undersea cable or dense wavelength division multiplexing (DWDM) equipment, the expansion of billing systems, and so on.
- Examples related to the introduction of new services include the addition of the toll-free (800) numbers, call forwarding, incoming call number identification, and so on.
- Examples on the effects of regulations on new telecommunication services include local number portability—that is, the capability to retain telephone numbers even after changing operators, the capability of locating the origin of an emergency call from a mobile phone, and so on.

Some projects may be related to the internal processes of the network such as changing to a more flexible billing system, providing on-line billing, and so on. For example, to retain its customers and increase their satisfaction, a service company may introduce a more granular rating engine to take into account fractions of a minute in the billing (the so-called "less-than-one minute billing"), even though this granularity would cause a direct reduction in revenue. In this case, the hope is that customer retention would offset the loss of revenues over the longer term [Goodpasture, 2002, p. 52].

Investment in public telecommunication falls in the same category of infrastructure development such as roads, railways, water and electricity and other public services. All of them are prerequisites for economic development and for improving the quality of life. Telecommunication services are also important because they complement other structures for communication such as postal service and travel. Given the three main dimensions of a project (time, cost and quality or scope), the goals of projects in public networks are typically in the domain of quality of service. The main factors that govern projects in public networks are [United Nations, 1987]:

- Level and distribution of income
- Industrial and technological level
- Economic growth rates
- Demographic structure (population, size, age structure, number of households, patterns of urbanization, mobility, etc.)
- Regulation
- Culture

Establishment of Specialized Business Networks

Private networks are used by enterprises or government entities for their internal communication. A private network can be also used by a federation of enterprises such as the networks used that link auto manufacturers with their suppliers. The scope of some of these projects may be restricted to a given building or campus, but may also encompass a federation of enterprises. For example, the Society for Worldwide Interbank Financial Telecommunications (SWIFT)—established in 1987 by 239 banks in 15 countries—has its own private network to relay the interbank messages related to international fund transfers.

The notion of the "tipping of network coalition" due to Professor Eli Noam provides a good way to explain the relation between public and private networks. A telecommunication network is a cost-sharing arrangement among several users to meet their communication needs. Initially, external subsidies sustain the growth of the network until it becomes large enough to attract subscribers willing to join to benefit from the networking effect because the cost per subscriber decreases as their number increases. At a certain network size, however, some potential users will add more cost than their contribution to the value of the networking arrangement, because their specific requirements are not economic to meet (e.g., remote locations, peculiar security arrangements, etc.). When this happens, the network expansion stops and—provided that the technology is ready and the regulations are favorable—those who could not join will band together to form other networking associations [Noam, 1992, pp. 26–42]. This explains why cost is the main consideration in private networks unless they transport mission-critical traffic, in which case quality remains the most valuable attribute.

Private networks are useful when the industry is organized in a tiered fashion such as the global automotive industry, which is dominated by a small number of integrating firms (General Motors, Ford, DaimlerChrysler, etc.) and a three-tier chain of suppliers as shown in Figure 1.3. The hundreds of suppliers that form the first tier use the products from the second tier, which numbers around 5000. The third tier consists of about 50,000 suppliers. As a consequence, the European automobile manufacturers have established a network called ODETTE for the exchange of information between suppliers and car manufacturers. Similarly, the Automotive Network eXchange (ANX®) is the network of the Automotive Industry Action Group (AAIG) (http://www.aiag.org) to link auto manufacturers with their suppliers in the United States.

The ANX is a virtual network in the sense that each participant can manage their part of the network and that several suppliers participate in building the connectivity. As a

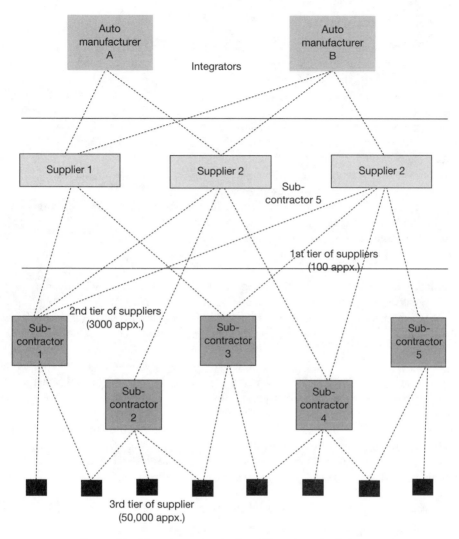

Figure 1.3   Three-layer organization of the automotive industry.

consequence, successful delivery of the service to the end-user relies on the collaboration of many types of service providers. Their equipment and their methods and procedures must be compatible—for example, procedures for maintenance and trouble-shooting of defects and outages. The infrastructure providers are responsible for managing the physical bandwidth for transmission (fiber cables, undersea cables, satellites, etc.). The various network providers build, operate, and maintain the network elements and infrastructure. To secure the communication, all users are certified and have a digital encryption certificate to protect the exchanges using the Internet Protocol Security IPSEC. Various certification authorities manage the encryption infrastructure, while an overseer orchestrates the overall functioning. Telcordia Technologies (formerly Bellcore), a subsidiary of SAIC, has that role while Verisign manages the responsibility for the security. The various digital certificates must be interoperable.

Temporary Networks

Temporary telecommunication installations consist of several networks for voice, data, or video associated with specific events such as major international conferences, global sporting events (e.g., the World Cup for soccer), or relief operations. The basic constraint in these projects is that the network must be operational at a date that is absolutely fixed. This means that it is possible to change either the scope of the project, the quality of the operation, the cost of the operation, or all of them. In the Athens Olympic Games of 2004, the Tetra (Terrestrial Trunked Radio) network for the security forces (police, coastguard, and emergency services) had to be installed sufficiently in advance to allow for the training of the security forces. To meet that date, Motorola had to assume the risk of rolling out the network before the contract was signed [Hope, 2003].

The size of such networks can be very huge. During the Hajj season (pilgrimage of Muslims to Mecca), the network load reached about 1.3 million telephone calls per day in 2001, with about half of them placed over international lines. For the Olympics Games, the typical call load hovers around 12 million telephone calls. International communications increase by about 20–30%, with most of the increase after the opening ceremony. It is similarly observed that the cellular phone usage usually peaks during the opening ceremony. Peaks to individual countries depend on gold medal performance and can exceed 100% of the total capacity. For illustration, Table 1.1 contains statistics for the evolution of the logistics needed during successive Summer Olympics between 1988 and 2000 [Verveer, 2001].

The main purpose of temporary networks is to provide access to other networks and to furnish timely information to the participants in the event in question (e.g., weather information, press conferences, meetings, press information systems, etc.) through a variety of access points (fixed-wire, mobile, radio, TV, satellite, etc.) and to connect them to the outside world. These projects have an absolute end-date that must be met at any cost (including sacrificing some functionalities).

The project tasks cover planning, installation, deployment, operation, and, finally, dismantling. Installation of temporary networks includes defining the following:

1. A numbering and addressing plan.
2. A frequency plan for radio, satellite, or wireless communications.
3. A network plan with redundancy through multiple routes, dual homing to avoid single point of failures, power backups, and so on.
4. Management of operations including customer support through call centers and messaging centers, network care systems to detect troubles, and the integration with

Table 1.1 Evolution of the Overall Logistics for the Summer Olympics 1988–2000 [Verveer, 2001]

	Seoul 1988	Barcelona 1992	Atlanta 1996	Sydney 2000
Athletes:	9,627	9,905	10,630	11,116
Press:	4,930	4,880	5,954	5,300
Radio and TV:	10,360	11,433	13,954	14,292
Ticket sold:	3,306,000	3,812,000	8,384,290	7,000,000
Number of events:	237	257	271	300
Countries broadcasting:	160	193	214	220
Number of sports:	23	25	26	28

other national networks and international networks (emergency, hospital and police services, broadcast, etc.). This may also involve arrangement with local operators to house the equipment needed for the temporary network at existing offices or relay sites.

5. Procedures for accounting, charging and billing.
6. Construction of storage sites, cabling, installation of antennas, and so on.
7. Physical and network security as well as disaster recovery plans.
8. Training of personnel and availability of necessary supplies for maintenance and repair.

CHARACTERISTICS OF TELECOMMUNICATION SERVICE PROJECTS

From the previous examples, it is seen that telecommunication service projects consist of a portfolio of subprojects characterized by the following:

1. Complexity of the interfaces (internally and externally)
2. International orientation
3. Multidisciplinarity
4. No mass production
5. Diversity of user requirements
6. Relatively long planning stage (even for temporary installations)

Complex Interfaces

The complexity of telecommunication operations has recently increased because service companies are no longer vertically integrated, have outsourced many functions, and rely on numerous intermediaries in their delivery process. Furthermore, new products or services must be backward compatible with the legacy systems and the prevailing organizational arrangements. This is true not only for new services in traditional telephony but also for mobile or IP applications, because they have to interwork with the existing fixed-wire networks.

To facilitate discussion, we consider two types of interfaces: (a) external interfaces to other entities participating in the service delivery and (b) internal interfaces connecting the processes and the operational systems of the operator.

External Interfaces. Figure 1.3 shows the increase in the number interfaces in the current architecture of telecommunication services. Typically, there is a network provider of record from which the interconnection service is purchased and which assumes all customer-facing support functions including sales, services, billing, and so on. Cooperation among the network operators can vary from interconnection agreements to allow transport and delivery of customer's traffic, to telehousing of equipment, to a full service agency. In this case, the operators may participate in the pre-sales discussions, in the ordering and provisioning of the product infrastructure, and in the deployment of the necessary network elements and management systems. For example, in the installation of undersea cables, teams from several companies and countries collaborate on the specifications, the selection of equipment vendors, and the definition of the network architecture. They also establish the financial and accounting procedures among the various partners and the procedures to be used for the procurement, installation, testing, and commissioning of terminal equipment.

A typical service provider assumes more roles than a reseller. A resale situation is a marketing arrangement whereby the reseller acts as the intermediary between the customer and the service provider; that is, it assumes most or all of the customer-facing functions (sales, billing, collection, etc,). One reason for such an arrangement would be to extend customer services to a different linguistic group or to overcome sale force limitation. A service provider, in contrast, fulfills some basic functional needs that the unbundling of telecommunication services has uncovered by bringing to the surface issues that were once internal to each telecommunications operator. Consider, for illustration, a disaster recovery service. Here, the service provider offers recovery facilities by replicating the customer's data centers and rents the necessary infrastructure from the network provider. The network provider, in turn, designs a network configuration with preassigned (but inactive) backup ports and access circuits for each customer's data center. The infrastructure provider may own the access circuits. However, it is the end-customer that designates which circuits will be activated to ensure that mission critical applications are minimally affected by the failure. Activation of the disaster recovery plan is triggered when the customer reports to the service provider a site failure and requests reconfiguration. The disaster recovery service provider, in turn, calls the infrastructure network provider to effect the change.

Thus, in a world defined by technology change, unbundling of services, and deregulations, in addition to the ambitions to achieve global connectivity as quickly as possible, hybrid service arrangements abound, with parts provided by any different number of suppliers using many technologies.

Internal Interfaces. Sales of telecommunications service begin with an initial customer contact or inquiry and conclude with a signed contract and the hand-off to the ordering and provisioning organizations. This process involves the customer, the sales team, and the capacity management team. The processes and systems used to support service establishment for public data networks, which are illustrated in Figure 1.4, are very similar to those used in traditional telephony [Rey, 1983, p. 374].

It is possible to group the functions needed for service delivery into six aspects [Ward, 1998, pp. 97–100]:

1. *Acquisitions and Sales.* This is a function that addresses all the activities associated with the acquisition of new customers. It includes lead generation, prequalification, proposal development, pricing, and contract preparation.

2. *Order Entry/Order Handling.* This is a function that includes the tasks associated with converting the request for service into a firm order. This includes finalizing the design details including firm order confirmation and order tracking and archiving.

3. *Provisioning and Installation.* These processes depend on systems for configuration and inventory management. The complexities associated with the handling of inventories in a large network that encompasses a variety of equipment are usually underestimated.

4. *Network Management and Trouble Management.* These are processes that rely on systems that are only accessible for the network operators for fault management

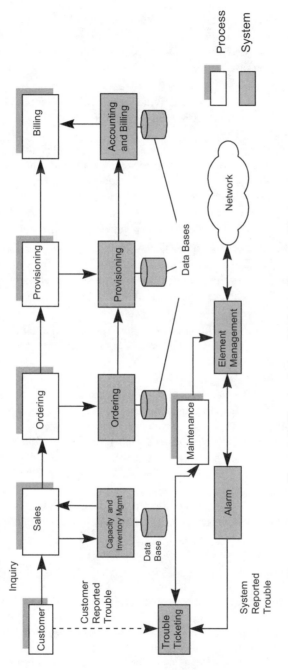

Figure 1.4 The classical model of operation support systems.

13

and performance management, including alarm monitoring and maintenance systems. These systems originate from many vendors, which may be problematic, because they are often vendor- and equipment-specific. This lack of standardization is problematic as vendors merge, drop products, or combine product lines. The integration of these systems for network management becomes incumbent on the network operator, which is a costly endeavor. Furthermore, if the end-to-end connection involves several administrative domains, such as on international links, the exchange of trouble reports or tickets and accounting data among network providers and their customers becomes complicated.

5. *Accounting and Billing.* These activities relate to the collection of the data elements for invoicing the customers. Mediation systems poll the network elements such as switches and pass the records to the billing systems. One important element of billing is the rating engine that applies the business rules associated with a service to compute the amount to be billed from the data in the call records. The computation takes into consideration a complex set of parameters such as the type of traffic (voice, data, text, etc.), the accounting rates set in the contract, any promotions or discounts, and so on. In the case of services spanning several countries, the currency to use is also a factor. If electronic payments are used, the security of payments and the privacy of data must be ensured.

6. *Fulfillment.* This is another process (not shown in the figure) that is related to any post-sale or post-installation customer activities, such as training, notifications of upgrades, service enhancement, and so on.

More on Billing Systems. Billing systems are crucial for the long-term survival of the network operator not only by recovering the cost of offering the service but also for uncovering new business opportunities. Customer relationship management (CRM) systems, for example, rely to a large extent on billing records to understand the profile of the existing users and how to tailor the services to better fit their needs or to attract new users. Unfortunately, many business managers underestimate the volume and the complexity of the data going through these systems, especially given the plethora of existing and new service offers that technology changes and changing market arrangements impose on a network provider that would like to survive in a competitive environment.

In a typical communication, subscribers are located on different networks, and thus multiple operators participate in handling the call end-to-end. Interconnect billing systems keep track of the traffic flowing between two different network operators so that the amount that they have to pay each other is consolidated in a net sum that is paid by a single bill.

In the United States, the Federal Communications Commission (FCC) mandates that revenues from long-distance communications be shared among all the networks that have carried the traffic as an exchange for the use of their facilities used to provide the call. Both the access billing process itself and the tools for tracking and calculating the bills are known as the carrier access billing system (CABS).

Finally, revenue assurance software helps the network operators sift though billing records to detect the source of revenue loss (or leakage) from inaccurate or incomplete records due to system malfunction, operator's errors (particularly due to subscriber churn), incorrect rating, hacking, fraud, and so on. This is important because the revenue losses can vary from to 2% to 5% and can reach as much as 15%. Losses are particularly high when new services are offered or when providers consolidate their billing systems following a merger.

International Orientation

In theory, manufacturers have the option of focusing on specific markets without having global ambitions. This is not possible in the case of communication services because they do not stop at geographical or political boundaries. Even if a telecommunications company decides to restrict itself to a region, be it a remote or a rural area, it would still have to receive traffic from, and send it to, the rest of the world. The combination of privatization and deregulation have added international flavors to investment decisions in telecommunication services, particularly because business globalization has shown the importance for "global carriers"—that is, carriers active in different regions of the world to serve the telecommunications needs of multinational corporations. At the present time, the service lead-time (i.e., the waiting period before the service is available) is not uniform throughout and depends on many factors. Some of these factors are the status of the local network, the access to the network including the way to negotiate with the venues owners for access, and installation of the necessary equipment and cabling. Another problem is that different operators have different policies for trouble reports and may use different formats for trouble tickets. Therefore, it is not possible to have a uniform policy for trouble detection and resolution in a worldwide enterprise network. A third inconvenience is the operation of help desks because of variations in holidays, vacations, time zones, work weeks, languages, worker's rights, and so on.

For a service provider, some of the challenges facing a seamless global operation are as follows:

1. The regulatory procedures vary from country to country with respect to licensing requirements, spectrum-management, environment impact, rules on the location of antennas or cell towers, the enforcement of service level agreements, the homologation of equipment or individual cards, and so on.

2. Variation in the performance obligations, particularly for voice communications such as dial tone delay, connection to emergency services, communications available to the deaf, and so on.

3. Differences in legal systems concerning rights and obligations of content ownership, responsibility of the carrier with respect to the content, right to privacy, encryption, and so on.

4. Account settlement and payment with many different currencies pose challenges for budgeting purposes due to the fluctuations in the exchange rate as well as changes in taxation laws. The global carriers will identify cost components per country for taxation purposes and devise ways to split the overall bills into the correct local currency for each country. This is important if the carrier offer network-wide consolidated discount schemes.

Multidisciplinarity

Implementations of telecommunication services involve several engineering disciplines (construction, physical design, mechanical, thermal, electrical, computer science, etc.) in addition to statisticians, marketing and legal professionals, and so on. Many of these aspects are intertwined. For example, the construction of buildings and the installation of antennas must be fire- and earthquake-resistant. Environmental regulations control the placement of transmission towers to protect the population from the radiation while the installation of satellite antennas must take into account resistance to wind. Risk analysis and disaster recovery rely on a combination of engineering, financial, and legal expertise. The operation and maintenance of the network require administrative skills for accounting, logistics, human resource management, and so on. An interesting example happened with the TAT-8, the first fiber-optic transatlantic cable. When the glow from the cable attracted sharks attacks, the failure rate increased unexpectedly. To resolve the problem, the opinion of marine biologists was solicited to design a suitable shield to prevent these attacks.

No Mass Production

Most telecommunication services take place in a specific environment within specific procedures and technical constraints. As a consequence, no two networks are alike because they have to fit within the environment that is defined by the legacy of previous technical and business decisions, legal rulings, and the operator's history. Thus, each project has a different context (politically and technologically), user population, and so on, even if the technology is well understood such as in the case of the plain old telephone service (POTS), private line, or traditional data services (e.g., frame relay). Variations in telecommunication projects depend on several factors such as:

- The type of network used (public, private, virtual private, etc.)
- Target market (consumers, business, government, military, service resellers, etc.)
- Nature of the installation (permanent or temporary)
- Types of service (voice, highly reliable data, best effort data, integrated traffic)
- The geography of the areas of the project; this is especially important in mobile networks where the topography of the environment affects the propagation of the signals
- Legal framework

One consequence of this characteristic is that the boundary between the "end of the project" and the beginning of production and life-cycle management is less defined than in the case of equipment design and production, especially for in-house projects. In such a case, the project development team may be called for some field support in case of problems, particularly in the case of testing the repair ("patch" testing).

Diverse Users

Success of the service project depends on the level of customer satisfaction with what was delivered and how it was delivered. Yet users are not homogeneous but fall in different subgroups, each with different membership needs. In an enterprise, the success of a service depends on whether it has helped the functional organizations in improving their functions. In a temporary project for a conference or a sports event, the needs of the participants are different from those of the media organizations. In mission-critical applications the reliability of the network and its availability cannot be compromised. However, there are many other tolerant applications where "best effort" is good enough.

A Relatively Long Planning Stage

Planning for telecommunication services usually takes a long time. Even temporary projects (e.g., disaster recovery contingencies) require a lot of planning and preparation (5 years in the case of the Olympics, for example). The planning includes aspects related to the network and the equipment, the organizational arrangement, the regulation, training of the personnel, whether permanent or temporary, and so on. The logistics of equipment removal is important, not only for temporary installations where typically more than 90% of the installed equipment has to be removed, but also in more permanent installations due to more stringent environmental laws that require recycling or the treatment of pollutants.

The industry of telecommunication services does not experience short turnaround times such as those encountered in consumer electronics, desktop software, or even network equipment: While the industry is dynamic, service innovations happen at a much slower rate [Ward, 1998, p. 38]. Evolution must be steady according to a thoroughly thought-out plan, even though the regulatory changes, globalization, and new technologies are bringing open significant opportunities. This fundamental point was missed during the dot com bubble.

Finally, the long gestation of telecommunication services makes them true reflections of the society itself in terms of fundamental assumptions and power structure: whether services are decided or financed after consultation with the public, by top-down fiat or through the market mechanism of supply and demand.

Summary of Distinctions Between the Development of Telecommunication Services and Equipment. The production and delivery of telecommunication service usually comprises many subprojects that evolve at different speeds, using a wide variety of technologies and requiring many distinctive skills. These challenges make telecommunication projects very rewarding because of the many possibilities of cross-education and the lack of monotony. Table 1.2 summarizes the distinctions between development projects in telecommunication services and those in product design and manufacturing or software development projects. These distinctions are important to ponder and keep in mind because they guide the way resources and skills are aligned during the lifetimes of projects. It is recommended that these differences be considered in the conception, design and implementation of telecommunication services.

SUMMARY

Telecommunication service projects are complex endeavors that exhibit two facets, depending on the vantage point. With deregulation, the number of interfaces and linkages as well as the pressure to achieve faster rate of returns have increased the complexity of service development. We drew on some examples to illustrate how the constraints on schedule, cost, and quality vary with the nature of the service. In particular, the main

Table 1.2 Dichotomy of Telecommunication Projects in Equipment Manufacturing and in Service Delivery

Item	Telecommunication Equipment Manufacturer	Telecommunication Service Provider
Stakeholders in the multifunctional team	Manufacturing, marketing, research and development, environmental	Administrative, legal, construction, quality assurance, marketing, environmental
Mass production	Yes	No
Procurement	Individual components or subsystems	Equipment, bandwidth, or other telecom services
Customer	Distributors and end-users	Depends on the business model (customer or business)
Clear boundary between project termination and life-cycle management	Yes	No, particularly for in-house development
International dimension	Optional—Mostly for marketing, regulatory aspects and compatibility through standards	Marketing, regulation, interconnectivity, account settlement, payment, troubles isolation and repair, vendor support, etc.
Quality criteria	Cost, size or footprint, power consumption, reliability, ease of repair, etc.	Cost, availability, reliability, billing accuracy, customer support, end-to-end quality, etc.

constraint on public services relates to quality; in enterprise environments, cost is the major concern while timeliness is ahead of all other considerations for temporary installations. We have identified six characteristics that distinguish the development of telecommunication services from the corresponding activities in equipment manufacturing. The rest of the book will demonstrate how this dichotomy affects project implementation.

2

STANDARDS AND INNOVATION IN TELECOMMUNICATION SERVICES

Telecommunication projects build on a technical infrastructure to satisfy business and so-cial objectives. In this chapter, we take advantage of studies on the management of inno-vation to gain insight into the nature of telecommunication projects. In particular, we ex-plain the role that internal and external standards play in the development of telecommunication services. The methodology presented here will combine two perspec-tives, the technological and the marketing and social, to draw a more comprehensive view of the context in which projects are executed.

THE TWO DIMENSIONS OF TELECOMMUNICATION PROJECTS

The Technological Dimension

There are five main stages of a technology life cycle: innovation, improvement, maturity, substitution, and obsolescence [Betz, 1993; Khalil, 2000]. These stages are shown in Fig-ure 2.1, with the ordinate measuring the market presence (e.g., revenues, market share, etc.).

An emerging technology stimulates the consolidation of new functional areas and the accumulation of new types of knowledge through research and field experience. As the properties of this emerging technology become better understood, new designs ameliorate its performance and increase the efficiency of the production processes. If the technology moves to the main stream, its market share expands until its performance saturates. At this point, any substantial performance improvement will require a switch to a new tech-

Managing Projects in Telecommunication Services. By Mostafa Hashem Sherif
Copyright © 2006 The Institute of Electrical and Electronics Engineers, Inc.

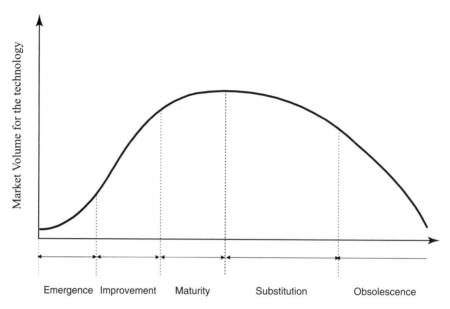

Figure 2.1   Technology life cycle.

nology as shown in Figure 2.2. It is sometimes possible to predict these limits using scientific knowledge, such as in the case of Moore's law in integrated circuit design or the channel capacity in information transmission. Some examples of past and current technology transitions in telecommunications transitions include:

- Move from manual switching, to automatic switching, which allowed direct dialing of telephone numbers
- Move from analog to digital in transmission and switching equipment, which allowed improvements in the voice quality suppressing noisy backgrounds
- Move from circuit-switched networks to packet-switched networks
- Move from coaxial cables to fiber-optic cables, etc.

We give now two examples showing the technology S-curve (but do not illustrate the technology transition shown in Figure 2.2). The first example is that of dynamic routing in circuit-switched networks, which was introduced in telephone networks to improve the efficiency of network usage and enhance robustness to failures [Ash and Chemouil, 2004]. The genesis of the work started in 1975 to 1980 by adapting the techniques of learning automata to routing problems for telephone. Figure 2.3 shows the diffusion of this technique from 1984 until 2000 in various telephone networks. The second example, illustrated in Figure 2.4, is the subscriber growth of i-mode services in Japan from February 1999 until November 2004.

The sequence of events in which technological change affect telecommunication services follow the following sequence:

1. A new technology is embodied in the design of network elements and/or network element management systems.

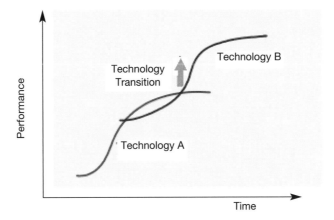

Figure 2.2   S-curves for two successive technologies.

2. Depending on the degree of novelty, the structure of the development process, including the production system, is modified.
3. The skills are transferred to the workforce of the equipment manufacturer and the telecommunication operations (engineers, technicians, managers, support personnel).
4. The supply chain may have to be modified (vendors selection process, acceptance test procedures, intervals, etc.).
5. Additional capital expenditure have to be approved (new facilities, test and diagnosis equipment, etc.).
6. The knowledge and experience is diffused toward the general public so that new applications are discovered and developed.

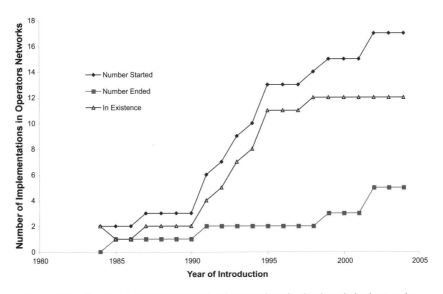

Figure 2.3   Growth of dynamic routing implementations in circuit-switched networks.

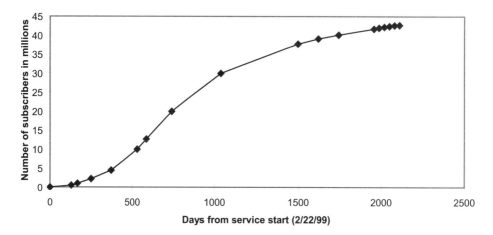

Figure 2.4 Subscriber growth for i-mode in Japan (Source: http://www.nttdocomo.com/company-info/subscriber.html).

In all these steps, successful management of the technological innovation depends on the mobilization of a network of people with technical and managerial skills. Members of this network include research and development specialists, production engineers, managers capable of facing uncertainties (entrepreneurs), and financiers. Because most of the know-how resides in the heads of key people, the transfer of knowledge depends on the movement of these key people and a systematic program for training at all levels. However, in the early phases, the networks are susceptible to single points of failures if a key participant with essential expertise leaves the project.

The Marketing and Social Dimension

C. M. Christensen introduced the concept of a value network as the set of attributes used to rank products, services, or technologies and determine their cost structures [Christensen, 1997, p. 32, 39–41]. This value network defines the context within which a company evaluates the environment that surrounds it, responds to opportunities and threats, and strives for profit. Significant changes in the attributes or their rankings that alter the industrial structure perturb that understanding by introducing discontinuities in the value chain. Factors that can cause a discontinuity include new legislation, emerging standards, evolution of the customer's profiles, and so on; such a discontinuity opens opportunities to new entrants. For example, how to evaluate the subjective quality of speech communication depends on whether mobility is important to the user; in cases where it is important, then some degradation can be accepted [Johannesson, 1997]. Innovations that change the rank order are called *disruptive* while those that preserve it are called *sustaining* [Christensen, 1997, p. 39].

The marketing impact of a change in the value network can effect one or several of the following aspects [Abernathy and Clark, 1985]:

1. Customer groups and markets
2. Customer applications
3. Channels of distribution and service delivery

4. Customer knowledge

5. Modes of communication with customers

Classification of Innovations

Depending on the degree of changes they introduce in the technology or in the existing value network, innovations can be grouped into four categories as shown in Figure 2.5: incremental, architectural, platform, and radical [Abernathy and Clark, 1985, Betz, 1993, p. 394; Sherif, 2003a, 2003b].

Incremental (or process or modular) innovations build upon well-known technological capabilities to enhance an existing technology through improved performance, enhanced security, better quality, and reduced cost, within the established value network. The purpose of the innovation is to enhance the competitive position through economies of scale to lower cost and improve productivity through automation. The objective of reduction in production and distribution costs requires extensive data collected from real experience. It is estimated that half of the economic benefit of a new technology comes from process improvements after the technology has been commercially established [Christensen, 1997, p. 56, note 3]. This is why incremental innovations are typically process innovations that tend to reinforce the existing industrial order because they are more readily integrated within the firm's strategy from both the technological and financial viewpoints. This contrasts with other types of innovations that could alter the order and offer opportunities to new entrants [Betz, 1993, p. 369]. For example, the high-level data link protocol (HDLC) is an incremental innovation through the standardization of the synchronous data link control (SDLC) protocol that IBM had developed for its System Network Architecture (SNA).

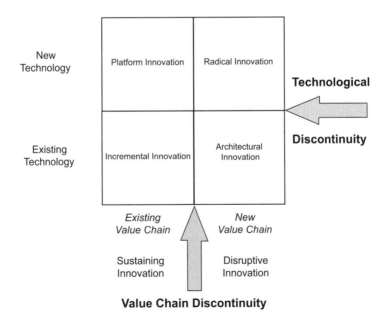

Figure 2.5 Classification of innovation in terms of the value chain and the technological competencies.

Architectural innovations (sometimes called systems innovations) provide new functional capabilities by redefining the rearrangements of existing technology to satisfy unmet needs (simplicity, cost, reliability, efficiency, convenience, etc.) [Betz, 1993, p. 21]. Architectural innovations result from a market pull: new uses of an existing technology. This is an articulation of latent patterns of demands that can be satisfied by blending incremental technical improvements from several previously separate fields of technology to create a new product or service. This category of innovation tends to modify the supply chain and to reorganize the market segments, ultimately forming a new value network [Christensen, 1997, pp. 171–176]. An example of architectural innovation is the automobile that put together carriage technology, with bicycle technology and the new gasoline engine.

The interplay between architectural and incremental innovations can be seen using the following example. The Standardized General Markup Language (SGML) was an incremental innovation from ISO starting with GML (Generalized Markup Language), a language that IBM had developed to manage electronic documents. The HyperText Markup Language (HTML) and the Extensible Markup Language (XML) descend from SGML. HTML is an architectural innovation based on SGML because its field of application is document retrieval over the Internet, which is in a different value chain than the original application of data base management. Finally, XML is an incremental innovation of HTML. This is summarized in Figure 2.6 [Egyedi and Loeffen, 2001].

Figure 2.7 illustrates the succession of innovation types from GML to XML using the 2 × 2 matrix of innovations.

Platform innovations correspond to a quantum leap in performance without changes to an existing value chain [Betz, 1993, pp. 309–322; Christensen, 1997, pp. 62–63]. This technology transition demands the integration of sophisticated resources and the exploitation of expertise gained usually beyond the reach of small- or medium-sized companies [Christensen, 1997, p. 74, no. 3]. New platforms change the technical competitive positions and weaken small firms by changing the technical characteristics on which competition is pursued. Because a technology push is the main characteristic of platform innovations, technological considerations dictate business strategies to manage the diffusion growth including licensing, training, and so on.

Radical innovations provide a totally new set of functional capabilities that are discontinuous with the existing technological capabilities or value networks. Kuhn [1970] showed that the advancement in science is characterized by long periods of regular developments punctuated by periods of revolutions. Likewise, radical innovations are spaced in time and, when successful, lead to a dominant design that is improved continuously.

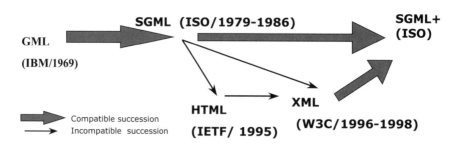

Figure 2.6 Evolution of GML to SGML, HTML, and XML.

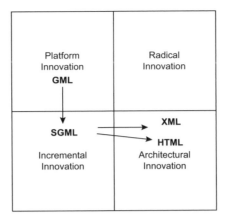

Figure 2.7   Relation of incremental and architectural innovations in the case of GML, SGML, HTML, and XML.

Radical innovations face four types of uncertainties: technical uncertainty, resource uncertainty, organizational uncertainty and market uncertainty. Technical uncertainty arises from two factors: (1) Many of the technical characteristics of the innovation are not well understood, and (2) an even better technology may become available and displace the technology under development. In telecommunications, this is how optical transmission displaced the emerging waveguide technology.* Resource uncertainties relate to the unknowns regarding the cost of development and implementation as well as of maintaining the collaborative network of technical, managerial, and marketing experts. Organizational uncertainties are due to the tension from simultaneous discontinuities in the technology and in the value network. Market acceptance is another unknown, because the more radical the technical innovation, the less likely that existing customers will be able to guide its development: Market research methodologies typically focus on existing applications [Betz, 1993, p. 165; Christensen, 1997]. For this type of invention, the advice by W. E. Deming is very appropriate: "New products and new types of service are generated, not by asking customers, but by knowledge, imagination, innovation, risk, trial and error on the part of the producer, backed by enough capital to develop the product or service and to stay in business during the lean months of introduction" [Deming, 1986, p. 182].

Innovations and the Technology Life Cycle

Consider Figure 2.8, which relates the various innovation types to the technology life cycle and market acceptance. Radical innovations are encountered first in proof of concept implementations. Once the innovation proves itself, successive platform and incremental innovations enhance the performance and allow the firm to gain market share. Architectural innovation are common when the technology has matured in pursuit of new markets. New product or service concepts based on disruptive innovations, whether radical or architectural, are difficult to envision. They require flexibility to unlearn habits and practices, carry some experimentation, and exhibit keen awareness of the environment in terms of competitors, suppliers, regulations, fashions and fads, and so on.

*I am grateful to my colleague Thomas Hellstern for suggesting this example.

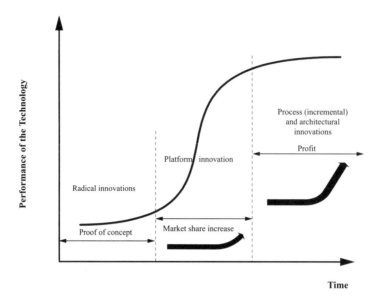

Figure 2.8   Relation of the types of innovations to the technology life cycle.

When the performance improvement levels off, process (incremental) innovations increase the efficiency of operations or enhance some features to increase revenues. Incremental innovations target existing users; typical customers' surveys provide useful guidance of their needs. Thus, incremental innovations follow a well-defined path from research to development, manufacturing, and deployment. During each step, the responsibilities are well-defined and the execution follows well-honed procedures. Thus, these innovations depend on the preservation, reproduction, and maintenance of past data, learning from past experience and specialized knowledge, skills, and capabilities.

INNOVATION IN TELECOMMUNICATION SERVICES

Services offered on public telecommunication networks are available to subscribers sharing a common infrastructure within the province of a network operator. To consider innovations in public telecommunication services, the whole system has to be partitioned into relatively independent modules that can be analyzed more or less independently. Radical innovations introduce new networking technologies that require new OSSs and new M&Ps and are often associated with new user applications. The need for new OSSs arises because a new set of parameters track the operation of the new network. Therefore, to monitor the performance, to detect troubles and localize faults, and to maintain inventories, new tracking systems need to be developed.

Platform innovations consist of improved networking technologies, improved OSSs, and improved applications. Incremental innovations build on mature network technologies with enhancements to the M&Ps and/or the applications. Finally, architectural innovations depend on sustaining networking technologies, improved OSSs, and new applications. Figure 2.9 summarizes the various categories of innovations in public

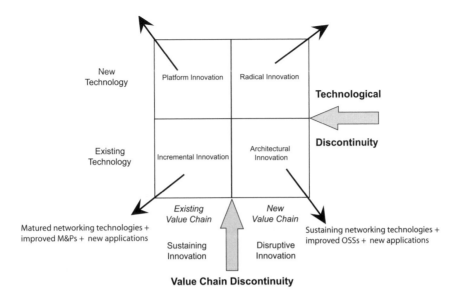

Figure 2.9   Innovations types in telecommunications services.

telecommunication services [Sherif, 2003a]. The changes in OSS for incremental innovations are not shown because they are on a much smaller scale than for all other types of innovations.

Incremental Innovation

Incremental innovations revolve around the dominant design to improve operational performance, reduce cost, and increase efficiency. Because the technical and operational properties of the service are well understood, these innovations depend on the empirical experience gained from extensive use of mature technologies and within the prevailing order. From a service viewpoint, the telephone answering machine is one such an innovation [Vercoulen and Van Wegberg, 1999]; without affecting the structure of the public telephone network, it increased the operators' revenues because callers left their messages to the absent party rather than hanging up, thereby giving the operators the opportunity of recuperating the cost of the call attempts. For mobile operators, games are incremental innovative services that use the same infrastructure to increase connect time. Within the core network, incremental innovations are improvements to achieve higher volumes and attain economies of scale. This can be realized, for example, with (a) increases in transmission or switching capacities by going to higher speeds or by using larger switches or (b) increases in process capacity to raise productivity through automation, relocation to areas of low labor cost, or outsourcing of jobs.

As mentioned earlier, the main organizational characteristic of incremental innovations is that they take place in a stable environment, where it is relatively easy to identify improvements with methodical planning. Most stakeholders have fewer incentives to disrupt the value chain, so these innovations tend to support the established order or dominant design configuration. Kuhn's description of the way "normal science" operates gives us some insight on the limits put on incremental innovations: They should be consistent with the established design [Kuhn, 1970, p. 39]; that is, they should not rock the boat. By optimizing the operation, however, they increase the rigidity of processes and products, which decreases the overall ability to cope with changing markets or technical requirements. Also, such productive and efficient operation is less robust to changes in regulations or customer tastes, breakdown in automatic processes, or disruptions in the outsourced functions.

Architectural Innovation

Many telecommunications services are architecture innovations. The reverse charging (800) service changed the way telephone calls are paid for—that is, by the called party instead of the calling party. Bluetooth (and the IEEE 802.x standards) is a marriage of local area networks (LAN) and wireless communications [Keil, 2002]. Bluetooth is a wireless technology that operates in the unlicensed part of the radio spectrum reserved for industrial scientific medicine (ISM) band at 2.45 GHz to connect mobile handsets with computer terminals. It relies on expertise in radio chip integration in addition to radio transmission, antenna design, and protocol engineering to communicate with portable computers and personal digital assistants.

The international call-back service was an architectural innovation to allow cheaper overseas phone calls. Similarly, the Simbox innovation exploits the price differential between calls made from fixed phones to mobile phones and those among mobile terminals. A box of SIM (Subscriber identification module) cards is placed on the premises of an enterprise so that calls from fixed phones to mobile phones are intercepted and rerouted over the mobile network using one of the available SIM cards.

Another example of architectural innovation is the i-mode service whose "father," Keiichi Enoki, admitted that it was made by combining existing technologies [Nakamoto, 2001]. This service concept provides wireless access through mobile telephone handsets to information servers with the public network operator acting as an intermediary to guarantee both the merchant and the buyer and to collect payments on the merchant's behalf. In other words, this is a new embodiment of videotext services that the minitel had offered in the 1980s with wireline access. As shown in Figure 2.10, the videotext technolo-

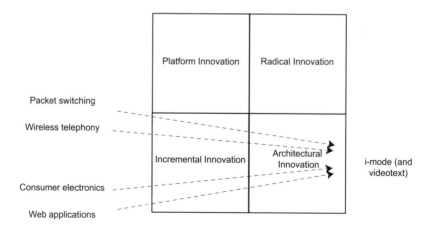

Figure 2.10 Videotext (I-mode/minitel) as architectural innovations for telecommunications services.

gy is based on a combination of display technology, computer telecommunications, and telephony.

Figure 2.11 illustrates how home systems (domotics) combine the technologies of wireless local area networks with appliance engineering and consumer electronics to add new functionalities to existing applications.

From a service viewpoint, camera phones are also an example of architectural innovation that add mobile telephony and handset ergonomics to the technology mix of digital photography such as image processing, miniaturized optics, and digital storage. By doing so, they changed the supply chains for handsets as well as for digital cameras and created security issues for corporations and institutions. Bundling Wi-Fi services with cellular telephony is another architectural innovation. Wi-Fi technology started to appear in airports, hotels, and cafés as web-connected hotspots to provide wideband access as an alternative to fixed connections using DSL (digital subscriber line) systems or cable modems. Similarly, location-based services are combinations of mobile telecommunications with location technologies. Finally, web-enabled commerce is based on numerous architectural innovations. In all these services, there is a change in the value chain. Location-based services, for example, can enhance business opportunities but raise moral and legal questions on privacy and security: Tracking a user's movements provides an indication of their absence from their residences.

Clearly, market pull is the main characteristic of architectural innovations. Their scope is to find new combinations of existing building blocks (e.g., technology, marketing channels, processes, etc.) to expand the market by satisfying unarticulated needs that are not met by existing services. This is why decentralization, deregulation, and the opening up of markets stimulate mostly architectural innovations. When the number of potential players increases, improvements can extend to areas that were not even considered because of limited resources (time, money, personnel, etc.). When quick returns are expected, however, the entrepreneurial efforts will concentrate on service innovations that combine existing technologies rather than on breakthrough activities.

Because architecture innovations are improvements based on the existing dominant design (i.e., no breakthroughs in technologies), it is important to correctly identify which innovations are architectural and which are incremental extensions of the existing design,

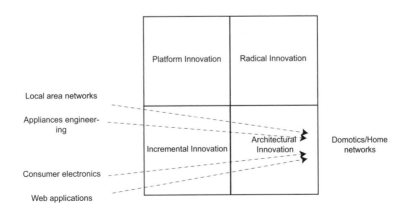

Figure 2.11 Domotics/Home networks as an architectural innovation.

to direct technical contributions in the right direction: whether to push the performance limits or to think outside the boundaries of the existing framework.

Platform Innovation

Platform innovations improve the performance of the telecommunication service without disturbing the industry structure. These are complex programs that require large capital investments to upgrade the existing infrastructure. Because of this, they weaken the relative positions of small firms due to the scarcity of financing as well as technical and managerial talent. The introduction of digital transmission in the 1970s was a platform innovation. According to the classification scheme in Figure 2.9, frame relay and the asynchronous transfer mode (ATM) are platform innovations of packet switching along the "connection-oriented" paradigm. Gigabit Ethernet can be viewed as either a platform innovation (from a local area network viewpoint) or an architectural innovation (from a wide area network viewpoint).

Radical Innovation

According to Kuhn [1970, p. 92], "a scientific revolution is a noncummulative developmental episode in which an older paradigm is replaced in whole or in part by an incompatible new one." If we consider that paradigms of telecommunication services are defined by both the technology and the market context through value chains, we can appreciate that radical innovations are infrequent because they change the way telecommunications problems are understood and resolved. The introduction of modems in the late 1950s was a radical innovation because it allowed the transmission of digital data among computers over the analog telephone networks. Another radical innovation is packet switching, which branched into two approaches. The "connection-oriented" approach continued along the lines used in telephony so as to maintain the overall call quality. The "connection-less" approach of the IP protocol, however, requires a major overhaul of the OSSs, new rules for traffic management, and retraining of human resources. This is why government intervention is often an important facilitator for radical innovations in telecommunications.

Once a radical innovation becomes the dominant design, continuous improvements in the form of platform and incremental innovations move the technology on its S-curve trajectory. However, architectural innovations can combine it with another value network, making it a potentially disruptive technology for another industry [Christensen, 1997, p. 41].

Interaction of Innovations in Equipment and Services

End-to-end service offers in telecommunications rely on the smooth integration of many components. This is why the classification scheme illustrated in Figure 2.9 considers process and business aspects in addition to technology. One implication is that, because the evolution of systems, organizations, and technologies are rarely synchronized, innovations in terminals, network equipment, and services are not necessarily of the same type, even when they appear as a single bundle to the end-user.

Earlier (section entitled "Incremental Innovation") we used the telephone answering machine as an example of incremental service innovation. From a terminal viewpoint, however, this innovation is architectural, because the terminal is a combined telephone receiver and a tape recorder.

Likewise, the view of cordless telephony depends on the vantage point. The design of the handset itself was an architectural innovation, even though the innovation was incremental for the plain old telephone service (POTS). Further enhancements to the terminal were of either the incremental or platform category without changing the status of POTS in any shape or form. For example, cordless telephony (CT) or digital enhanced cordless telecommunications (DECT) systems were two new platforms that the European Telecommunication Standards Institute (ETSI) standardized for providing wireless access to the telephone network for voice and data traffic in residences and business sites [Pandya, 2000, pp. 89–108].

There are also cases where innovations in equipment and services fall in the same category. Apple iPod and iTunes are architectural innovations in both the terminal side and the service side (downloading music and managing copyrights). Both are bundled together through a proprietary copyright management system.

PHASIC RELATION BETWEEN EQUIPMENT AND SERVICES

The following examples are given to show the phasic relation between equipment and service. Figure 2.12 shows the evolution of revenues for X.25 networking technology during its obsolescence. Clearly, the revenues from X.25 services exceeded those from equipment sales. Also, the peak of equipment revenues preceded that of service revenues by about 3 years in the United States and 6 years worldwide. The most probable explanation for the earlier peak in the United States is that the switch to frame relay was faster in the United States because X.25 public data networks were less common. While there is a huge discrepancy among the various sources on the magnitude of the worldwide revenues of X.25, both data sets exhibit the same trend. It is not possible to know with precision the reason for such a difference because the methods of analysis are proprietary. In any event, in real life, project decisions are rarely made with pristine or unequivocal data.

A similar trend can be observed with frame relay. Figure 2.13 depicts revenues from the sale of frame relay equipment and the revenues of services offered on public data net-

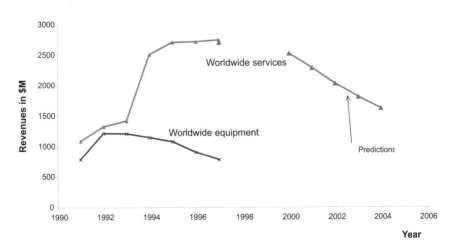

Figure 2.12   Revenues for X.25 networking technology during obsolescence (dotted lines represent extrapolated data).

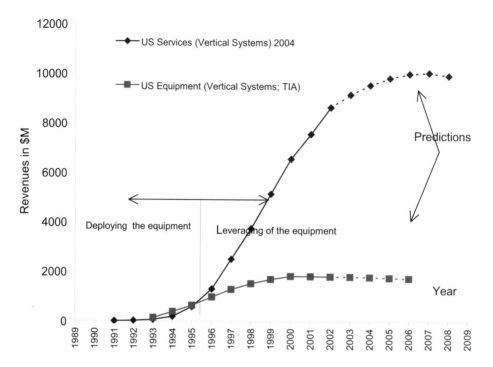

Figure 2.13 Equipment and service revenues for frame relay technology in the United States.

works in the United States. From 1995 onwards, service revenues exceeded equipment sales and continues to rise even though revenues from equipment saturated around 1999.

Figure 2.14 shows the worldwide revenues from ATM service and equipment as estimated in 2003 (later data were not available at the time of publication). The data indicate that revenues from public ATM services have not yet exceeded the revenues from equipment. In other words, operators have yet to leverage their investment.

Thus, for a given technology, the market dynamics for network equipment and for network services exhibit a clear pattern whereby the former may be considered a leading indicator of the latter. Similarly, the peak of the service revenues lags the peak of equipment sales by anywhere from 4 to 10 years [McCalla and Whitt, 2002; Sherif, 2003b]. This trend is sketched in Figure 2.15.

The time lag is the consequence of the amount of effort required to make a new service generally available on a public network, even when the value chain remains intact. Network elements need to be tested and deployed. This process includes vendor selection and a thorough laboratory evaluation that can last anywhere from 12 to 18 months for a high-quality reliable service. New network engineering rules have to be conceived and verified while the OSSs are modified to accommodate the new services. Maintenance personnel need to be trained and sales people prepared to explain the offers to potential customers. The magnitude of the up-front investment explains why, once the service is up and running, operators prefer to exploit it as long as they can. Furthermore, unless there are substantial savings (e.g., 10 times) in terms of cost or quality, customers prefer to avoid the disruptions of service migration. Clearly, Toffler's "economy of impermanence" [1971, p. 51] does not apply here.

The persistence of networking technology explains the long tail of the time series relat-

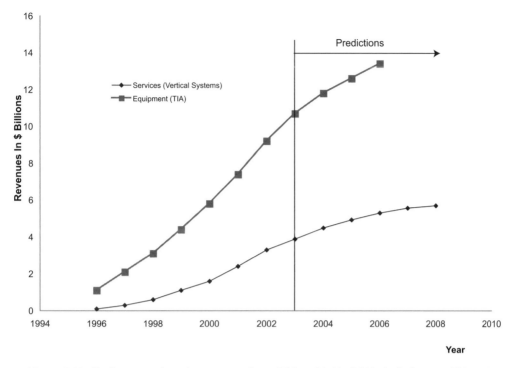

Figure 2.14 Equipment and service revenues from ATM worldwide (2003 view). *Sources:* TIA and Vertical Systems.

ed to service revenues. Assuming the most optimistic substitution scenario, along with using historical data on the diffusion of radical innovations through industry to estimate the duration of the transition, the substitution could take about 30 years with 50% of the substitution within 15 years [Betz, 1993, pp. 258–262]. This estimate is consistent with the lifetime of wide area network technologies, which is about 25 years. In fact, Forrester Research currently expects the shift to voice over IP to take about 14 years from 2006 to 2020 [Forrester Research, 2003].

Figure 2.16 illustrates two behaviors. First, we see a dramatic decline in the membership of the ATM Forum in parallel with the increase in equipment sales and service revenues in the United States. Clearly, the interest in a technology wanes after it is deployed in networks even though there are many issues that arise during the life of a technology and most of the benefits occur later in the technology life cycle. Furthermore, according to the available market predictions illustrated in this figure, revenues of ATM services in the United States will saturate before that of equipment sales and at a lower level. If these predictions are correct, U.S. network operators will be hard pressed to recuperate their investment on the ATM infrastructure that they have deployed.

STANDARDIZATION FOR TELECOMMUNICATION SERVICES

Public telecommunications services require specific agreements among the parties involved (equipment manufacturers, network operators, service providers, and end-users).

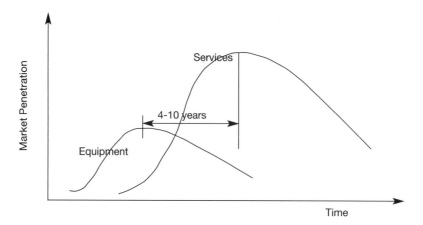

Figure 2.15 Platform innovations for telecommunication equipment and services.

The diminishing role of governments in mandating the rules for telecommunications means that these agreements will be based on voluntary standards—that is, on the business strategies of the various firms that form the industry.

Inter-firm standardization in telecommunications has focused on the network elements and their subcomponents because the outlay for standardization is easier to justify in this case. It is known that the value of a network increases with the number of its users—the so-called network externalities. The absence of a common interface standard is a burden on all parties. For example, having a dual digital transmission standard (the (μ-law at 1544 kbit/s in North America and Japan with the A-law at 2048 kbit/s for the rest of the world) adds a step to all digital transmission equipment. Similarly, the fragmentation of

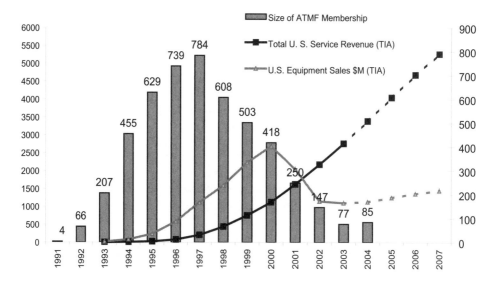

Figure 2.16 Evolution of the membership in the ATM Forum and the sales of ATM equipment and services in the United States.

the market for cellular telephony in the United States into islands of competing digital standards [IS-95, IS-136, GSM (*Groupe Spécial Mobile*) as well as Nextel/Motorola's proprietary systems] increases the cost of interconnection and prevents the consolidation of the industry.

A telecommunication service company has additional interests in external standards. Some of these reasons are as follows:

1. Reduce the uncertainties concerning the compatibility of equipment from different manufacturers.
2. Establish a framework for bilateral and multilateral negotiations among carriers. The emergence of virtual network operators that buy capacity from network providers highlights the need for such standards.
3. Ensure continuity of the equipment supply.
4. Avoid the monopolistic power of a single source.
5. Synthesize knowledge in the form of rules thereby reducing the dependence on rare technical expertise.

Standardized routines within each service company define the various methods and procedures used to eliminate the idiosyncrasies of relying on subjective judgments, which is exceptionally important in the case of emergencies or disaster recovery. Selection of the course of action should be part of the discipline of project management. One possible tool to help make that decision is to consider the timing of standards, which is the subject of the next section.

Timing of Standards

By relating the type of standards to the technology S-curve, it may be also possible to evaluate whether standardization makes any sense. One consideration is the timing of the standard, which can be considered on the basis of either (a) the marketing needs of a specific product or (b) the evolution of the technology itself [Sherif, 2001].

Marketing Perspective. With the introduction of a product or a service in mind, standards can be *anticipatory, enabling* (*participatory*), or *responsive,* as shown in Figure 2.17. This model gives us an appreciation of the relationship of the standard to the market development—that is, time to market, time to scale, and time to profitability. The cycle starts when the need for a product or service arises irrespective of whether the anticipatory standard is in place or would have to be developed.

- *Anticipatory standards* are essential for the widespread acceptance of the product or service.
- *Enabling (participatory) standards* proceed in lock step with the introduction of the product or service.
- *Responsive standards* codify a product or service present in the market or define the expected quality of a service and performance level of the technology.

The advantage of matching the standards to the intrinsic capabilities of the technology is that it gives us a realistic assessment of standardization at a given juncture of the technology life cycle. This is explained next.

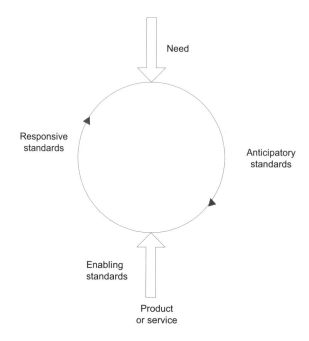

Figure 2.17 Standardization within the product or service life cycle.

Technological View of Standards. By relating the type of standards to the technology S-curve, it may be also possible to evaluate whether the standardization of a technology is taking into account the improvement of the performance of the technology that the end-user perceive [Betz, 1993, p. 308]. The relationship depicted in Figure 2.18 between the innovation type and standardization shows that anticipatory standards correspond mostly to proof of concept activities. Enabling standards, in contrast, are typically associated with performance improvement while responsive standards correspond to process innovations.

Note that the quality of a telecommunication service results from the collective effect of the performances at several layers. Assessment of the performance involves a mixture of subjective and objective parameters to assess the quality of transmission in the presence of impairments. It covers the user interface to the service offers, the network performance from the aspects of switching and transmission, and the operational aspects and the service support functions [CCITT, 1993; Oodan et al., 2003].

Anticipatory Standards

Anticipatory standards are crucial for successful interoperation of communication systems. Their specification runs parallel to the production of prototypes, to pilot services, and to field trials. They also provide a way of sharing ideas through a systematic way of distilling investigations and experimental data into useful engineering knowledge. This is useful when the risks are high because the collaboration with other competitors working on the sets of problems can increase the chances of success. The incentives to standardize in this phase are less when one organization is so far ahead of the others. However, in the case of radical innovations, standardization may also help legitimize the new technology.

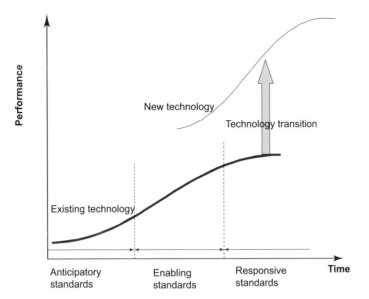

Figure 2.18   Standardization and the technology S-curve.

The Wireless Access Protocol (WAP) is a specification for Internet access through mobile terminals defined by a consortium of companies (i.e., it is an anticipatory standard). This was an architectural innovation to support of a large number of available and planned wireless protocols and a variety of portable terminals (palm-top, phone, computer, personal digital assistant) with a variant of HTML—Wireless Markup Language (WML)—suitable for the display data on mobile terminals. Its commercial failure shows the importance of a correct reading of the market pull.

Some earlier examples of anticipatory standards are the X.25 packet interface, ISDN (Integrated Services Digital Network), TCP/IP, the Secure Sockets Layer (SSL) for secure end-to-end transactions—its standardized version TLS (Transport Layer Security) is an incremental innovation—Bluetooth, IEEE 802.11, and so on. The UMTS (Universal Mobile Telecommunications Service) is an anticipatory standard for new services for voice and data (144 to 344 Kb/s outdoors, 2 Mb/s indoors). Thus, the transition to the third-generation wireless may be viewed as a discontinuity in the value chain and in the technology competencies. It is still not clear what services would attract interest and what configuration of customers, competitors, and suppliers will emerge.

Anticipatory standards are susceptible to misguided efforts whenever field experience is lacking and when the market requirements are unclear or ill-defined. For example, the Open System Interconnection (OSI) transport and management protocols became hopelessly entangled in attempts to satisfy the requirements of all parties involved. Also, sticking to wrong market assumptions may lead to a dead end, such as in the case of Group 4 facsimile (facsimile on ISDN). Ideally then, anticipatory standards should offer a minimum set of features to stimulate the market for services and, with feedback from pilot studies, define the production environment of the new technology. Such a restricted scope reduces the chance of overspecifications that could lead to onerous implementations. It avoids premature commercial conflicts that can stall the standardization, provided that it is clear what aspects are expected to evolve and when they will evolve.

Enabling (Participatory) Standards

The definition of enabling standards proceeds in parallel with market growth and enhancements to the technology and the products. Thus, enabling standards describe refinements in the production system, in the product systems that embody the technology, and in the application systems.

The advantage of enabling standards is that they diffuse technical knowledge and prevent market fragmentation. Without an enabling standard, the risk of incompatible approaches that could result in a fragmented market becomes important. The manufacturers whose technical members are leading the effort to specify the standards must consider the cost of developing the standards and the risks of market fragmentation. They may have to release technical information earlier than anticipated to conduct the necessary tests. As further changes and refinements are made to the draft standards, the manufacturers and service providers must review their products periodically to bring them into conformity with the new standards. Finally, enabling standards signal that product differentiation is shifting to areas not covered by the standards (cost, quality of implementation, service support, etc.).

The GSM specifications constitute a standard for digital mobile telephony because they defined a platform for future growth both for service operators and for manufacturers. Once this platform was established, network operators grew through acquisitions* that called for architecture innovations to integrate the different operations support systems of individual operators.

Enabling standardization helps establish and improve the dominant design. An example is that of V.90 recommendations for modems at 56 Kbit/s. There were some proprietary designs of modems operating at that rate; however, to avoid market fragmentation as well as increase the overall market size, various manufacturers agreed to collaborate in the standardization process at the ITU to come up with an implementation that would work independently of modem manufacturer.

Once the dominant design is formed, responsive standards are needed, including OSS interface standards.

Responsive Standards

Once the dominant design has stabilized, organizations embed practices into routines to interpret the information and retain only the most relevant information stream. Responsive standards relate to the routine manifestations of the technology in a service system or precursor products [Betz, 1993, p. 361]. They codify knowledge gained from past successes and solutions with the goal of facilitating the task of problem detection and resolution in the future.

For example, GPRS (General Packet Radio Service) and SMS (Short Message Service) are responsive standards from the point of view of service provides in the sense that the market was already established and the performance of the GSM system had reached a saturation level. In contrast, manufacturers would tend to see these standards as anticipatory because they are needed for the design and deployment of the equipment. Clearly, there is some tension between manufacturers and service providers in terms of the urgency of coming up with a specification. Other examples of responsive standards are the

---

*Vodaphone, for example, acquired Airtouch in the United States in 1999 and Mannesman in Germany in 2000.

various methods for the evaluation of voice quality through objective and subjective means, the measures for the overall quality of services, and so on. In the IT world, the traditional role of ISO TC97 was to improve existing specifications and turn them into international standards. As explained earlier, SGML was built on GML and HDLC on SDLC.

Lack of Standards

There is more temptation to avoid standardization and rely on proprietary protocols in architectural innovations, making them of the "walled garden variety." This is illustrated by many cases such as the minitel/i-mode, Apple's iPod, Bloomberg's terminals for real-time financial information, and the smart card, even though they use standardized components to take advantage of the benefits of mass production in terms of cost, availability, and so on. The major reason is that architectural innovations are combinations of existing components so the real barrier to competitors is not the technical expertise but the switching cost of users. Thus, it is not easy to transfer a collection of music encoded in Apple's digital rights management software to a new format. Similarly, a finance manager would like to avoid the cost in time and effort of switching to another information provider and relearn unique sets of commands and keystrokes to access real-time financial information [Gapper, 2004]. However, the benefits of network effects may encourage manufacturers to standardize to increase the market size as shown in the case of Bluetooth and WAP.

Even though information technologies are embedded in a wide variety of telecommunication products and services, the standardization for OSS interfaces is lagging behind that of network equipment interfaces. This has several consequences. First, the communication channels for the on-site workforce (OSWF) that operates and maintains services across several administrative domains (e.g., on international links) remain primarily the phone, fax, and e-mail. Obviously, this is an error-prone process that can increase the total cost of operation by around 10%, as shown by the justification data for EDI (Electronic Data Interchange). Trouble detection and isolation is even more problematic because it requires coordinated manual activities. Another consequence of this lack of standardization is the expense needed to update or develop new OSSs for each new service without being able to reuse what is existing for other service offers. As a result, a typical large operator would end up operating and maintaining a few hundred of such systems for all the various services and networks that it operates and maintains. Perhaps one of the lessons from the failure of so many joint ventures or mergers among telecommunication operators [Curwen, 1999] is the need to include the time and effort to consolidate the distinct OSS's or at least make them communicate with each other in the scope of the merger.

Many are now realizing that OSS standardization facilitates back-office integration and reduces the cost and time to roll out any new technology in the network. This explains the various "zero provisioning" initiatives to increase the efficiency of provisioning through automation of the relevant processes.

To facilitate the standardization of OSS interfaces, IT providers, given their traditional reluctance to open interfaces, have to be convinced that this is in their best interest. OSS standardization can be then presented as an opportunity to provide operators with efficient solutions for operations and maintenance once they have built their networks and reduced their equipment purchases. The main selling point would be that most of the benefits from platform innovations take place later in the technology life cycle.

Standards Policy and Knowledge Management

Knowledge management is the process with the following objectives:

1. Spread the knowledge so that it is not restricted to specific people.
2. Generalize local knowledge in a form that can be used in a generic way.
3. Provide a framework for discussion and exchange of information among all those involved in the activities.

Knowledge management in telecommunications services covers three areas: technology performance, systems performance, and the end-to-end service performance. Some aspects are specific to the firm because they give it some competitive advantage. They will be covered in the chapter on communications management. Aspects that are general to the industry as a whole and are common across several firms are typically within the purview of standardization activities. In particular, anticipatory standards are used for knowledge creation by defining a framework for subsequent developments. Enabling standards correspond to the diffusion of knowledge and to the large-scale adoption of the framework.

Usually, it takes anywhere from a year (in the best case) to 3–4 years (on the average) and up to 8–10 years (in complex systems) for an external standard to be developed and successfully implemented in commercial products. Firms approach standardization in several ways that can be listed from the least responsive to the more responsive as follows:

1. Following the dominant trend. This approach uses the least of the firm's resources at the expense of two major risks: missing the chance to improve the standard or influence its development and misinterpreting the intent of the standard.
2. Passive participation in the standards bodies, to collect the necessary documents and understand the context of the standard.
3. Technical contribution toward a standard. Contribution toward a standard implies a policy of knowledge management—that is, that of generating, keeping, and/or releasing information.
4. Building or joining alliances to influence market expectations.
5. Starting a new consortium and attract other interested parties to approve or promote a given technology.

Selection of the course of action should be defined in the scope of the project. As was stated earlier, architectural innovations integrate many technological capabilities to meet the needs of new markets. Even if the service interfaces are kept proprietary, standardized components reduce the cost of operations and maintenance. This is a case where alliances provide the service provider with indirect influences over the standardization of the various components instead of active participation in each organization.

In the case of radical innovations, the typical decision is to wait until some prototype network element with the new technology is available. This approach uses the least amount of resources, provided that two major risks are avoided: (1) waiting too long until the chance of improving the standard development is lost, and (2) not having a good understanding of the technology leading to erroneous strategic decisions.

Typically, service providers alternate between passive and active participation in standardization activities, particularly in application and implementation consortia. They may also start specific consortia to deal with the applications relation to the operational and business applications such as the case of the TeleManagement Forum.

SUMMARY

Many organizational and technological trends that affect the telecommunications industry are bringing to the forefront the issue of service interoperability or maintenance in an environment of constant change. Technology and market dynamics influence projects of telecommunication services. This chapter presented a systematic way to assess the interactions between technology and the market context as they affect the service innovation and the type of standards that would be needed.

The main points are as follows:

1. There is a difference in outlook, time horizon, and interest between the manufacturers of network equipment and telecommunications service providers. This difference impacts the way they innovate and plan their standardization strategies.
2. Radical innovations in telecommunication services require government facilitation.
3. Deregulation stimulates architectural innovations, but these tend to be of the "walled garden" variety.
4. In a deregulated environment, however, standardization is essential for end-to-end service interoperability, particularly when there are numerous manufacturers and service providers.

3

THE PROJECT MANAGEMENT CONTEXT

In this chapter, we show how the work breakdown and the organizational breakdown structures can be used to divide the project tasks among functional groups and to constitute the project team. We then relate the organization of the project team to the type of innovation. Because a service company conducts many operational activities at the same time that it implements the project, we discuss how the rolling wave method for project management—also called phase management—can help balance the service portfolio to ensure the long-term operation of the service company. Finally, we show how the management by phases can be related to the build–operate–transfer (BOT) mode of delivering turnkey projects.

ORGANIZATION OF THE PROJECT TEAM

In every project, there is a core group and an extended support group. The core group consists of the team responsible for planning and executing all activities—that is, those stakeholders that are involved with project activities on a daily basis. The membership of the core group may vary over time, as people finish their task and move off the project to other activities. In this section, we discuss the various ways that this core group can be structured in ways to fit the characteristics of the innovation as presented in Chapter 2.

The first step in project organization is the definition of the work to be done. The *work breakdown structure* (WBS) is a product-oriented family tree of all the tasks needed to accomplish the project objectives. It provides a framework for tracking individual responsibilities and task assignments and the project progress. A perpendicular arrangement is the *organizational breakdown structure* (OBS), which is a functional family of tree of all functional groups that are needed to accomplish the statement of work. The project lies at the intersection of the two structures as shown in Figure 3.1: Both the WBS and the OBS are tools for managing the project. According to the ANSI/EIA standard EIA-748, the intersection point is called a control account. This is a point where management can exercise planning and control by assigning the work project to one responsible functional organization and by defining the project management responsibilities. Within each functional organization, the responsibility matrix indicates the project assignments of each member of the functional group.

The interactions among the functional and project control and communication structures affect the way the project is conducted. At a minimum, the project manager is responsible for developing the formal project plan, tracking the progress of the execution, and invoking the escalation procedures whenever needed. Additional responsibilities depend on the way the project sponsor divides the formal project responsibilities among the functional groups and recruits the project team. Accordingly, there are three main organizational types:

1. Functional organizations
2. Matrix organizations
3. Projectized organizations

Functional Organization

In a functional structure, skilled professionals are grouped into specialized departments. Day-to-day network operation and maintenance is a complex endeavor that requires discipline with strict rules and detailed procedures that constrain individual members within narrowly defined roles. This is why a typical organization in telecommunication services is arranged along functional lines as shown in Figure 3.2. The company has the typical support functions of any firm: human resources, administration, legal and finance, as well as marketing and sales. The departments that are specific to telecommunication services operate in the following areas:

1. Provisioning—that is, installation of the equipment and its configuration to connect with equipment or other networks for regional, national, or international communications.
2. Customer care to respond to end-users requests and trouble reports as well as for collecting the bills.
3. The network operations for regular maintenance, fault management, security, account processing, and so on.
4. The architecture and design group responsible for the technical planning of new service development, network evolution, and the evaluation of new technologies.

In a functional project organization, projects exist to support the functional mission, that is, their scope is defined by the boundaries of the day-to-day running of the business. Project teams follow the functional hierarchy and are bound by established sets of rules and procedures. In such an arrangement, there is a clear and precise delineation of roles and responsibilities: Functional managers supply the necessary resources and define the technical criteria (*how* and *where* the tasks will be done and by *whom*), and the project

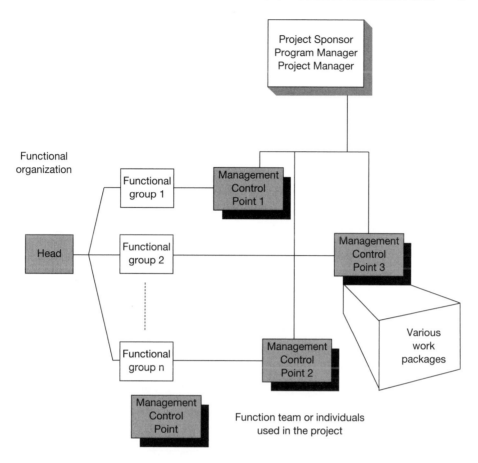

Figure 3.1 WBS/OBS integration in a functional organization.

manager defines the project constraints (*what* is to be done and *when*) [Kerzner, 1998, pp. 10–16] acting as a facilitator or a coordinator. Usually, the project manager does not have a clear or unequivocal authority to command the requisite resources or personnel yet is responsible for the outcome. Therefore, the project manager has to rely on the pressure of the bureaucracy to get things done while, at the same time, attempt to subvert it to face issues that are not "in the book." Using the terms of Herzberg's motivation theory, functional managers are responsible for the so-called hygiene factors (salary, status, job security, working conditions, etc.), while project managers are responsible for the motivators (definition of the work, recognition, career enhancement, etc.) [Dunn, 2001].

When a project cuts across several functional areas, there will be multiple subteams in each functional division. Once a subteam executes its part of the overall project, the output is handed to the next subteam for completion of the work. As shown in Figure 3.3, the coordination among the various functional areas falls in the sphere of action of the organization head.

Examples. The focus of day-to-day operations in a telecommunication service organization is the maintenance of network operations and the handling of requests for addition,

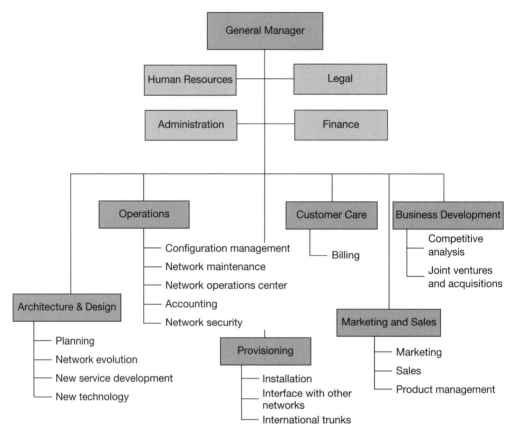

Figure 3.2 Functional organizations in a telecommunication services company.

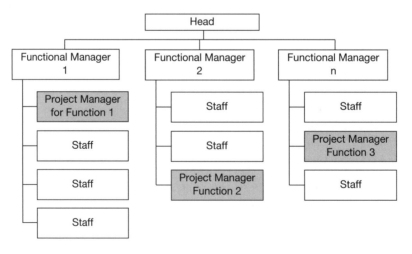

Figure 3.3 Functional organization of projects.

modification, or deletion of the configuration parameters to respond to subscribers' requests or to engineering changes. Periodically, the hardware and software in the network and its operation support systems (OSS) are modified through upgrades and/or replacements. As a consequence, most projects in telecommunication services are incremental in nature and are organized along functional lines. Some of these projects are:

1. Acceptance testing of specific hardware or software releases before deployment in the live network
2. Improvement of the quality of the OSSs, such as the capability to detect failures or to have remote diagnostics, and so on
3. Measurement of the end-to-end quality of service or benchmarking studies
4. Development of tools for bandwidth allocation planning

Advantages. Functional organizations are efficient because the tasks are subdivided to reach an optimum based on some criteria. They work well in stable environments. The main advantages of the functional organization of projects are as follows:

1. There is minimal disruption to existing operations.
2. Project responsibilities and functional responsibilities are aligned, which makes it easier to identify resources and evaluate their performance.
3. The experience gained enhances the collective functional expertise from project to project.

Disadvantages. The problem with functional arrangements is their propensity to maintain the status quo. The stability of the environment means that groups can remain unchanged for long periods, so that one or more of the following problems may appear:

1. The project manager may appear as an intruder that perturbs the daily routine.
2. The team members may fall into complacency and intellectual stagnation.
3. To maintain discipline and efficiency, the number of management layers has to increase. This leads to cross-functional communication difficulties, particularly when the control of information is leveraged in interdivisional rivalries.
4. Because project tasks are subdivided among the functional groups, it is not easy to get an idea of the overall project status at the working level.
5. In case of resource conflicts, functional responsibilities take precedence over project responsibilities. For example, in the case of a financial crunch, the functional organization would most probably prefer to focus on its own projects.

Matrix Organization

A matrix organization is a networked arrangement where project managers share some of the formal responsibilities of functional managers in the areas of allocation of resources and the evaluation of performance of team members. Figure 3.4 illustrates this project arrangement.

Depending on the balance of power between the functional organizations and the project organization, there are three types of matrix organizations:

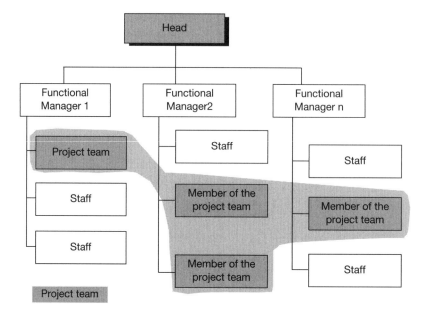

Figure 3.4   Matrix organization of a project team.

- A weak matrix (or "light weight") where the balance of formal power favors the functional manager with the project manager coordinating activities.
- A strong matrix ("heavy weight") where the balance of formal power tilts to the side of the project manager, while the functional manager supplies the resources.
- A balanced matrix, which is an arrangement where the project manager and the functional manager have equal formal and institutional power on the course of the project, because they are both at the same hierarchical level.

Figure 3.5 shows the arrangement of a balanced matrix organization. It goes without saying that the success of such a structure depends on the degree of communication and cooperation among the project manager and the various functional managers. When the team members are not co-located, a matrix organization is the default structure, even when it is not explicitly indicated as such. In such a case, communications will be stronger within each geographic area; therefore, key representatives of each area should be meeting, perhaps face-to-face, on a periodic basis to review the project status and discuss any technical difficulties that may have been encountered.

Examples

1. Introduction of a new software or hardware platform is typically carried with a balanced matrix organization that combines marketing and technical, with the technical team representing both the network elements and the network element management systems.
2. The management of network emergencies ("network meltdowns") and disaster recovery is typically arranged as a strong matrix.

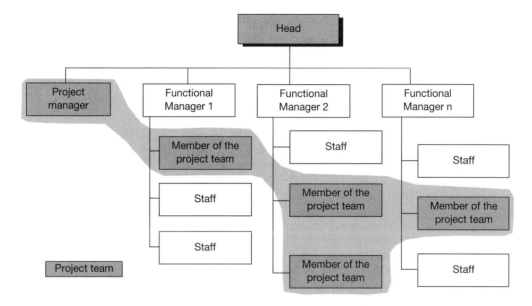

Figure 3.5   Structure of a balanced matrix organization.

3. The definition of a network activation date for a new business customer is usually done through a weak matrix structure, with the project manager coming from a marketing organization.
4. Migration of large enterprise networks, such as for insurance companies, from one platform to another mobilizes many functional areas as well as the customer's own project managers and the project managers of the various vendors. The private network may cover tens of thousands of locations ranging from corporate buildings to field offices as well as individual agent sites. At each site, the existing service will have to be disconnected and the new one connected. Quality of the operation depends on many minute steps such as validation of the physical addresses of the sites to be disconnected and reconnected, verification of the billing accuracy for the new service, resolving various exception cases such as orders with incomplete information or invalid addresses, customer requests for delays or rescheduling, and so on. Clearly this is a case of a balanced matrix (internally) and a weak one (externally).

When the project is large, the project is divided into many subprojects and the project manager becomes a program manager that overseas several functional project managers. These project managers coordinate activities with functional or regional groups. As an example, consider the case of the TAT12/TAT13 system, which was the first fiber-optic submarine "ring network" to link North American and Europe. The cable owners formed a consortium to share the construction and the maintenance of the cable systems for its expected life of 25 years. System upgrades are managed through a balanced matrix arrangement. One such upgrade took place to introduce the Dense Wavelength Division Multiplexing (DWDM) equipment that became available to increase the system capacity from 10 Gbit/s to 30 Gbit/s. The project manager was selected from one company and granted sufficient autonomy and authority to conclude effective negotiations with the suppliers and to supervise implementation of the project by the equipment vendors and/or by employees of the other members of the consortium. Some of these responsibilities were:

- Verification of the suppliers compliance with the contract conditions
- Tracking progress—in particular, importation and custom clearances of the equipment in the various countries
- Validation and approval off all billing milestones
- Management of the quality assurance testing
- Management of the project risks
- Provide written progress reports and make a quarterly personal presentation on the project status to representatives of the members of the consortium.

Advantages. Matrix organizations avoid many of the problems that functional arrangements face—particularly for nonincremental innovations. They improve the utilization of functional skills distributed across organizations and enhance the communication and coordination across functional silos. In the strong ("heavy weight") structure, the project manager has budget control over the project and can give input to the team member's performance review. This increases accountability of the project team in meeting the project goals because of lesser attachment to the functional objectives. This allows easier integration of the various elements needed for the "system solution."

Disadvantages. All matrix organizations disrupt the normal functional routines. In the weak ("light weight") structure, the project manager has little organization or formal control over people, budget, and so on. In the case of a balanced matrix project arrangements, the requirements from the project manager may conflict with those from the functional manager. This problem is acute in companies that operate in several continents. Consider the case of Global Crossing operation. Before it applied for Chapter 11 protection and eventually had to divest some of its companies, its main components were a U.S. company, a Japanese company, and a European company. In Europe, each of the major countries (France, Germany, the Netherlands, United Kingdom) had its own country organization, with the hub being located in the Netherlands. Thus, a project would be organized in the form of an n-dimensional matrix (or a "hypercube") by function, by region, by country, and by project. So a project team member working in a given European country will have to report to four different authorities as illustrated in Figure 3.6.

Projectized Organization

This type of organization is given different names: task force, tiger team, project, and so on. Basically, this is an autonomous team with individuals assembled from different functional areas to collaborate and produce an output that is more than the sum of their individual knowledge. Therefore, this is a learning organization, where the emphasis is on identifying obstacles to the goal and understanding their root causes to avoid them in the future.

To avoid the distractions from the day-to-day operations, the project team may be moved to a separate location under the direction of a project manager who was given all the necessary authority (financial, technical, etc.) to operate independently. Forming a dedicated unit is a characteristic of disruptive innovations, whether architectural or radi-

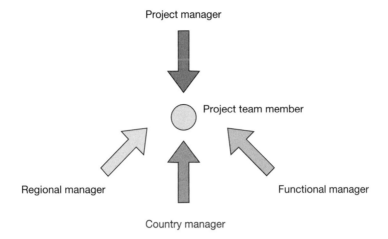

Figure 3.6 Pressure on the project team in a four-dimensional "hypercube" project organization.

cal. For architectural innovations, the aim is to encourage new integrative structures through reassembly of existing components in a new form. For project with high technical uncertainties such as in radical innovations, provided that the right amount of expertise is included, this structure increases the chance of ad hoc or spontaneous communication among the team members and allows the necessary control of resources and funding needed to push the envelope of knowledge and overcome technical barriers.

Examples

1. Preparing a response to a nonroutine or very large request for proposal (RFP).
2. Implementing a project for a customer. This is the classical form of that organization because the organization derives its revenues from performing projects for others (outsourcing).
3. Building a new network such as extension of a global network to a specific region
4. Acquisition and operation of a foreign company.

Advantages. The main advantage of this organization is that it encourages focusing the resources on the objective at hand and avoids the delays that would result from conflicts on resources between the functional and the project team.

Disadvantages. The first major disadvantage is that this arrangement tends to disrupt existing services. The second problem is more long-term. Having a separate unit both geographically and organizationally forces a historical discontinuity because the new team will form its own subculture that may not be in harmony with the culture of the original firm. This has to be considered if the team is to be reintegrated into the mother organization. Finally, when a projectized organization is used, each member will need to move on to something else, either to a new project or to a functional structure as the project ends. So, team members will have to spend some time on finding new positions. The transition is fraught with problems because the old organization may not have a position commensurate with the newly acquired expertise.

Comparison of Project Organizations

Project organization is not an end in itself. Its purpose is to integrate and align the re-sources available to meet the project goals with minimum transaction costs in decision-making and information distribution. Thus, organizational structures have a direct impact on the conduct of the project because they define the communication channels, the control structures, and the lines of authority. Table 3.1 illustrates the effect of the project structure on the role of the project manager, the technical lead, and on the way the performance of team members is evaluated.

PROJECT ORGANIZATION AND INNOVATION TYPE

The project management literature does not provide guidance on how to select the mode of project organization. A typical statement is that the "type of organization will depend on the "specific circumstances such as the technology or the people at hand" [AFITEP, 2000, p. 16]. It seems intuitive, however, that the structure of an organization handling an emerging technology or service concept should be different from the one for a growing field or when markets saturate or shrink. Fortunately, the taxonomy of innovations that was previously discussed suggests ways to associate the organizational structure with the innovation being implemented [Abernathy and Clark, 1985].

Incremental Innovation

Incremental innovations correspond to a stable design (the dominant design) that is being improved in small increments to enhance the quality and reduce cost. As long as the dom-inant design remains stable, knowledge accumulation takes place along functional lines in the form of standardized routines and procedures. This is why incremental innovations do not cause substantial changes to the organizational structure. This is the type of innova-tion that network operators tend to prefer because they can easily fit them without affect-ing network operation.

Feedback from users or customers identify ways to make improvements. As a result, organizations use their experience to filter information and extract what is relevant to its functional objectives. The price of this efficiency is that they may miss cues on changes in the environment or in their value chain. In addition, because of their investment in opti-

Table 3.1   Effects of Project Structure

Structure	Role of Project Manager	Technical Lead	Performance Evaluation of Team Members
Functional	Coordination of activities	Functional manager	Functional manager
Weak matrix	Definition of project plan	Functional manager	Functional manager
Strong matrix	Definition of project plan and selection of resources	Functional manager and project manager	Functional manager with input from project manager
Projectized	Definition of project plan and selection of resources	Project manager	Project manager

mizing their operation within the current framework, they become more conservative and interested in preserving the *status quo* [Henderson and Clark, 1990]. This explains the so-called "innovator's dilemma," where market leaders are not able to respond in a timely manner to emerging market trends that are changing the basis of competition in their respective industries. This inability is not related to technological deficiencies but to methods for performance measurements and resource allocations that systematically favor sustaining innovations over disruptive innovations [Christensen, 1997]. All these factors contribute to making functional organizations more suitable to incremental innovations that improve cost-effectiveness.

However, a stable organizational environment is characterized by sharp functional separations with centralized control as a coordination tool. Therefore, a goal should be to harness internal knowledge and expertise and integrate the dispersed knowledge among functional areas (e.g., from troubles in the field) and synthesize it in a form that can be used to drive incremental improvements

One way to break the fences among distinct functional organizations is to standardize the procedures using a uniform set of routines. In addition, standardization provides a reliable way to disseminate knowledge and transfer technology in structured way. Another possible way for knowledge diffusion can be achieved through the techniques of Total Quality Management (TQM) such as quality circles, project reviews, and code inspections.

The role of the technical team is to establish the limits of the existing dominant design. Incremental innovations are geared toward the existing paradigm, which can make the firm vulnerable to perturbations in the environment due to changes in the value chain. These changes can be due to one of several causes such as the legislation, technical advances, or users' preferences. It is important that the firm be ready to detect legitimate changes in the value chain rather than being oblivious to them or denying them.

Architectural Innovation

Architectural innovations result from the fusion of dispersed knowledge in response to a market pull that the collective intelligence interprets to blend threads from several previously separate fields of technology into a new product or service [Jin, 2001, pp. 186, 260]. Thus, the marketing organization has an essential role in shaping the innovation and defining its deployment schedule [Lapierre and Hénault, 1996]. Clearly, multifunctional teams in a projectized (or at least a strong matrix) configuration are most suited to nurture architectural innovation; sometimes the project is even spun-off the organization either physically or organizationally, to avoid the competition for resources with more established products. For example, the unit that developed the Bluetooth technology was geographically separated from the bulk of the Ericsson Mobile Communications (ECS) group.

The success of an architectural innovation depends on an imaginative evaluation of potential markets to find ways to reuse existing components of products or services and tailor them to meet needs. With the marketing people in the lead, the role of technical experts is to define the technical parameters that can be useful for the new marketing package and determine how much can the old rules of thumb be relaxed.

The case of Bluetooth shows some of the ways used to discover these new markets by combining the marketing forces of many parties. Ericsson built an alliance with chip manufacturers (Intel), other competitors in the mobile markets (Nokia, Motorola), and computer manufacturers (IBM, Toshiba), as well as with PC card developers in the form of

Bluetooth Special Interest Group, to promote the technology and define common interface specifications. Notice the absence of any service provider in this alliance because in architectural innovations, the market is usually not fully developed [Keil, 2002]. The interest of service providers in Bluetooth arose later, when using a mobile phone while driving became controversial out of public safety concerns. Bluetooth was then hastily incorporated in mobile phones to allow untethered hands-free communications, even though the security aspects of such an addition were totally understood.

Platform Innovation

Platform innovations happen mostly during the growth phase of a given technological configuration or a dominant design. Contrary to architectural innovations, they represent a technology push. New skills are needed and training is essential for successful operation of the network as well as for managing the new supply relationships. This is why the project manager must be a senior technical manager with solid technical background. Strong matrix organizations are most useful in the case of platform innovations because they facilitate the coordination among the various functional disciplines within the technical team. The aim of the organizational arrangement is to help the diffusion of the technical knowledge toward the marketing organization as well as toward internal and external users to apprise them of the technical and operational properties of the new technology, its limitations, and potential restrictions to its usage [Lapierre and Hénault, 1996].

Radical Innovation

Radical innovations require dedicated structures where the project manager has complete profit and loss responsibility in addition to the technical responsibility as well as the orchestration of the multifunctional activities. There is a need for insight and imagination into user's needs that can be met with the emerging technological capabilities. Projectized organizations are most appropriate to manage a small and tightly integrated multifunctional team that includes the necessary technical, managerial, and marketing skills with the same vision [Jin, 2001, pp. 189–203]. Breaking new ground implies that there will be a scarcity in technical, managerial, and marketing skills. Lack of experience with the technology or with the market can lead to unrealistic expectations about time, cost, and/or performance, particularly when the innovation is focused for a critical mission. Furthermore, user's requirements and profiles are not usually precisely defined, and there may be disagreements on the most appropriate way to meet them. This means that experimentation with its lot of mistakes and failures is part of the environment and that technology transfer and training is a key part of the activities. The project manager must provide the environment where creativity and experimentation are encouraged. However, to reduce the cost of such an experimentation, the size of the team allowed to perform that experimentation has to be limited.

Figure 3.7 summarizes the above discussion by relating the organization type to the technology life cycle and the innovation type.

The Role of the Project Sponsor

One important role of the project sponsor is to ensure that the organizational arrangement of the project team matches the innovation type and the life cycle of the technology. Signs

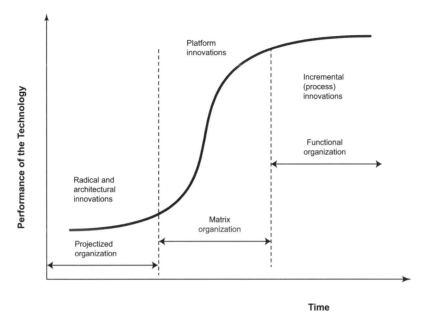

Figure 3.7 Relation of the project organization to the innovation type and the technology life cycle.

that a reorganization may be warranted include the following: (1) The system knowledge related to the innovation is lacking, (2) work distribution among one or several functional organizations is contentious and fractious, and (3) team members need to constantly switch among several regimes for knowledge management. However, the nature of the reorganization depends on the innovation type and the degree of the technical challenges that the project is facing.

The transition from a functional organization to a pure project organization increases the pressure on the whole organization to go beyond the narrow specializations and overcome organizational rigidities. This affects the smooth operation of existing services. Furthermore, there are many potential events that can change the business priorities, thereby affecting a project: funding constraints, technological uncertainties, organization conflicts, and so on. Also the reintegration of a team involved in a disruptive innovation has to be planned.

Once a project structure has been selected, some of the additional responsibilities that are incumbent on the project sponsor are as follows:

1. Articulation of a compelling vision; this is of particular importance for radical innovations.
2. Definition of the project scope in terms of business goals and measurable objectives.
3. Selection of the project leader that will be most effective in leading the type of project that is being considered.
4. Alignment of resources and cross-organizational priorities so that all organizations would treat the project with the same level of priority.

5. Delegation of sufficient authority to the project manager to lead the project. In that regard, the involvement by the sponsor in daily activities undermines the project manager's authority with respect to team members, suppliers, and customers.

6. Removal of organizational blocks that the project team may encounter as they arise.

7. Act as an escalation point when the project is placed in jeopardy and schedules cannot be met.

8. Act as a contact point for the team to the external world.

PHASE MANAGEMENT AND PORTFOLIO MANAGEMENT

A program is a group of projects managed in a coordinated way either because they are interlinked or because the benefits can happen when all the projects are finished simultaneously. Thus, telecommunication services projects are usually part of an overall program. At any time, there are many programs running concurrently, each of which consists of projects at various phases of execution and competing for a limited amount of available expertise and financial support.

Portfolio management is concerned with the overall positioning of the company and not individual projects. The objective of portfolio management is to balance investment decisions and ensure the longevity of the service company through a steady stream of revenues from a mix of products. These products are at different stages of their life; therefore, optimization of a technology portfolio requires an evaluation of the situation (financially and technologically as well as from the viewpoint of human factors). This entails adopting an integrated view, one that brings together issues that impact several functional divisions taking into account the capabilities available, the status of progress in each project, and the projected evolutions in technology and market needs. The review has to cover more than just financial figures and must be conducted with a deep understanding of the nature the innovations being considered to estimate the cost of development, the capabilities needed for the development, a realistic time for market introduction and for profitability, and so on. This is predicated, of course, on having a review team with the right mix of expertise and experience.

The Rolling Wave Method for Service Development

The introduction of a new telecommunications service consists of several phases [Ward, 1998, pp. 41–49]: opportunity analysis, definition and feasibility, network design, acceptance testing, controlled introduction, deployment, and general availability. Phase management is a structured way to provide independent nonadvocate reviews of the various projects and programs at a series of designated checkpoints for each project. At these points—called service gates or quality gates—a formal examination of the relationships and dependencies among the various projects and/or programs is conducted to verify that predetermined criteria have been met. Usually there are conflicting criteria at such decision points. Based on this analysis and using pre-established and measurable criteria, a decision is made as whether the project should proceed. This is why these points are called Go/No-Go decision points. If the project is to proceed, the plans for the next phase are then updated and refined based on the accomplishments so far, the availability of resources, and the changes in the environment or the starting assumptions. Thus, the phase management provides a framework to provide cross-functional guidance on the project execution. Figure 3.8 illustrates this iterative approach to service planning.

In the remainder of the presentation, we consider that the service development and roll-out will comprise the following phases:

1. Concept definition and feasibility evaluation
2. Initiation and preliminary planning
3. Implementation
4. Controlled introduction
5. General availability and project closeout

Phase 1: Concept Definition. In the case of disruptive innovations, the definition of the project boundaries depends on the results previously obtained from a feasibility analysis or a field trial to ensure that the expectations from the project are realistic and that all major requirements have been clearly defined. For sustaining innovations, market studies, customer input, and technical evaluations constitute the major inputs. Various techniques are available for cost estimation in the case of incremental innovations. For other types of innovations, the estimates are educated guesses that may or may not end to be valid. In any event, inputs from suppliers and subcontractors assist in this estimation. The output of this phase is a scope document, sometimes called a Market service description (MSD), that defines the basic assumptions in the project including:

1. The scope of the project and its objectives as defined in the MSD.
2. The authority levels, escalation procedures, and the roles and responsibilities of the project management team.
3. The processes and methodologies to manage the business case performance (data to be collected, how to track variances and their effect on the business case).

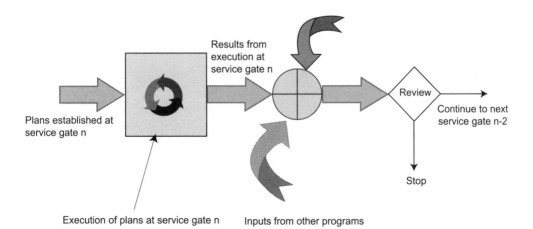

Figure 3.8 Iterative planning of a telecommunications project with the phase gate process.

4. The various decision points and criteria. For example, when and how to decide whether a product or service is ready for controlled introduction or for general availability.

5. The criteria to determine the client/customer satisfaction with the deliverable and how it is being delivered. These are the criteria that will be used to define the project performance and final success.

Phase 2: Initiation and Preliminary Planning Phase.   This phase starts in earnest once the project budget has been determined. The purpose of this phase is to save time and cost during project implementation by defining a baseline project plan. The plan specifies the final project goals, constraints, milestones, the structure and composition of the project team and its relationship with the functional organizations, a work breakdown structure, an organizational breakdown structure, various processes, tools and methodologies for tracking and change control, and so on.

During this phase, more precise interface and performance requirements are made to guide the evaluation and selection of technical solution. These requirements include the specification of services to be provided such as the geographic coverage, the service levels and quality, the billing arrangement, and so on. Except in the case of an incremental innovation, the technical plan should refrain from mandating a definitive technical solution at this stage. This is to avoid overconstraining the implementation team with an imposed and untested solution. Otherwise, if that implementation team discovers that a technical problem is a "show stopper," it will devote significant energy to convince the remaining stakeholders, rather than concentrating on progressing the implementation.

Phase 3: Implementation.   Implementation starts when the gatekeepers and the project team agree that the scope has been clearly defined and that it is possible to achieve the project goals (financial or otherwise). The process of implementation involves planning, executing, and control of the project tasks that are needed to deliver the project output. The implementation phase usually consists of three subphases: (a) detailed design, (b) laboratory testing and acceptance, and (c) installation, test, and service turn-up. The output of that phase is a defined architecture with specific equipment, the detailed equipment configurations, the geographic location of the network host sites, the required transmission capacities and procedures for spare storage, and returns of defective parts. Other activities include site survey and preparation, constructing, purchasing or leasing space for buildings, preparation of training material, training teams, and so on. Additional consideration may include establishing internal standards, participation in external standards bodies, and so on. Project management discipline is essential to track all the necessary threads.

Phase 4: Controlled Introduction.   Risk identification and quality of service can be improved by the use of controlled introduction of the product or service. In this phase, some intended users or customers under tightly supervised conditions use the service. Working in this way may identify additional risks that were missed in the initial risk identification effort and have a more accurate assessment of probability of occurrence of some risks. The performance of the network over extended operational periods is measured, and the service delivery procedures and customer satisfaction levels are evaluated and adjusted to assure that procedures and performance meet the requirements and service level agreement objectives.

In addition, the product or service is gradually integrated into the organization systems. Provisions are made to provide feedback to the software and hardware vendors if specific problems arise during field use. At the same time, the adequacy of the operations support systems and the methods and procedures are evaluated with an eye on future improvements.

Phase 5: General Availability and Closeout. General availability of the service to all geographic locations transfers the responsibility to organizations that manage the service throughout its life cycle. The operating organization must verify that the objectives agreed upon were met. Once this is done, most development resources return to their functional organizations for subsequent assignments; a subset is retained to support operation.

The closeout is the formal process of documenting the lessons learned for future reference and the closing of the business case. In a formal closeout, the actual schedule is compared to the baseline schedule. Similarly, the actual costs are compared to the budget, the quality of the deliverables is evaluated against the requirements, and the actual work delivered is compared to what was planned.

Increasingly, project management is including end-of-life planning; this is because of the push in many countries to recycle material and to treat hazardous waste. This ensures that documentation is maintained for the discontinuation of the service or product.

Canceling Projects. Project launches can stumble on two types of errors: the dogged pursuit of a failing project and the termination of a potential winner [Mankin, 2003]. One root cause for the first type of problems is a collective belief in the inevitability of success despite warning flags [Royer 2003a; 2003b]. This results from overconfidence of the project team or its deep aspiration for success or from an aggressive schedule that impels people to skip reviews and violate agreed-upon decision procedures. Another cause is overconfidence and the suppression of skepticism that nurture a cheerleading environment. Platform innovations, which are particularly susceptible to such a blind faith, can benefit from the establishment of unambiguous decision criteria and a careful monitoring of the project performance.

The second type of errors—canceling a potentially winning project—befalls disruptive innovations whenever bureaucratic inertia resists change, or when conflicts among decision-makers overwhelm the decision-making process. Companies that are not careful end up with the problems that C. M. Christensen has labeled the "innovator's dilemma" [Christensen, 1997] even though their origin is a refusal to admit the need for innovation.

Relation to the Build–Operate–Transfer Model

The build–operate–transfer model is common for turnkey projects where the project planning and implementation and initial operation are outsourced to a contractor. Phases 1 and 2 of the phase gate process—concept definition and high-level design—correspond to the *design* phase of the BOT model. One way to ease the transfer to the operator is to include some representatives from operations from the outset in the project team and not wait until operations start. This will allow consideration of the operational aspects during the design and construction phases of the project. Also, design decisions that will impact on the operations can be documented in a way that the operational team can understand to help them define the operating procedures for normal operations and during emergencies (network meltdowns).

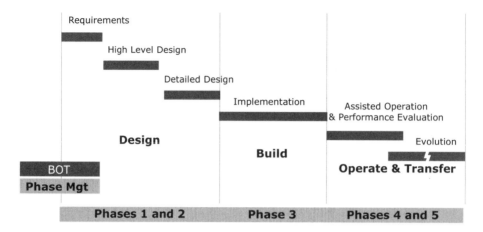

Figure 3.9   Relationship between the phase-gate process and the build–operate–transfer model.

The second major phase in the BOT model is *build* phase, where the network infrastructure and its supporting organizations and systems are put in place. This corresponds to the implementation of the network and organization and service test. The *operate and transfer* phase of the BOT model corresponds to the controlled introduction and the general availability of the service. In the BOT model, the service delivery organization gradually takes responsibility for day-to-day running of service (Figure 3.9).

SUMMARY

There is no unique structure that is suitable for all projects at all times. The main factor in the project organization is to facilitate the communication and collaboration communication among the project contributors and to provide the adequate balance among the functional entities. The choice of a structure depends on the importance of the project, its urgency, the type of innovation, the risks, and the project environment. Telecommunication projects are labor-intensive, and the project organization structure has to take into account the constraints on existing entities. One clear indicator that the current structure is inadequate is the presence of acute conflicts on resources. The organizational differences that are suited for each mode of innovation imply that a transition from one mode of innovation to another may pose significant organizational challenges to established firms. The quality gate process—also known as the stage-gate process—is a structured and systematic way to integrate the various subprojects and processes used to progress the development and deployment of an innovation. Reviews of the project status provide an opportunity for redefining the project boundaries, including the option of canceling unprofitable projects to release the resources other activities.

4

SCOPE MANAGEMENT

Scope management is the set of processes to ensure that the project delivers the work that was specified and to restrict the activities to what is needed for its completion with success. Thus, the first major task in a project is the definition of its scope—that is, its boundaries in terms of the features and functions of the product or service to be delivered (or retired), the schedule for delivery, and the budget. With the various dimensions of the work defined, planning for the project can start. Scope management is a systematic method to achieve constancy of purpose even when modifications are introduced during the project execution.

Typically, an executive sponsor outlines the project high-level objectives in the form of a *project charter*. The scope is further refined in a document that is often called a market service description (MSD) so that a work breakdown structure (WBS) can be derived and a project plan of record can be agreed to. A plan of record is the first step to ensure that all project participants share a collective vision regarding the project objectives, basic assumptions, and governance (authorities and responsibilities). As explained in Chapter 2, the gating process is a good way to manage these various phases without losing sight of the end goals. However, during the project execution, scope management focuses on keeping the project on track even though the project environment has evolved. Among the telltale signs of inadequate scope management are many reworks and "scope creep." This chapter presents two cases to illustrate how inadequate scope management can jeopardize telecommunication projects. The first example is about global telecommunication alliances. The second relates to Net 1000, a project of the 1980s to provide intelligent data networking.

SCOPE INITIATION

Project initiation is the set of activities that end with the definition of a project charter. These activities may be prompted by a need analysis, a feasibility study, a response to a customer request, and so on. Part of the purpose is to define the position of the product within the company's portfolio in terms of type of innovation and market potentials to identify potential opportunities and threats. In particular, market analysis can help focus on the triple constraints of cost/quality/time. New entrants, in particular, may depend more on outside research from market research companies because they lack customer contacts. Furthermore, because the success of architectural innovations depends on a market pull, marketing insights are essential to find the new needs that the service is expected to meet. However, in emerging fields or when market conditions are volatile, which are typical of disruptive innovations, the methodology of traditional market surveys may give biased or incorrect results.

The project sponsor, as a representative of the senior management of the company, initiates the project by issuing the project charter. The project charter is a brief summary of the contour of the project. It gives a general description of the scope, shows how the project objectives are congruent with the company strategy, identifies the project manager, and defines its authority. A project charter gives a general description of the project content and the project management structure and identifies the project sponsor and potentially the expected customers. In the case of telecommunications services, this phase requires an understanding of the possibilities of current and emerging technologies, new consumer trends, and potential competitors as well as existing and forthcoming rules and regulations.

SCOPE PLANNING

Capturing the right scope—not too narrow and no too wide—is not an easy task. It is possible to fall enamored with an innovation and to overestimate its capabilities. Or it is possible to look only at quick financial returns and avoid any investment for improvement. This is why scope planning is both an art and a science. It is an art because it requires judgment, wisdom, and courage to make decisions that satisfy all stakeholders. It is a science because there are tools that facilitate decision-making such as WBS. The trick is to match the outline in the project charter with the triple constraints of cost, schedule, and quality, the phase in the life cycle of the technology, and the type of innovation (incremental, architectural, platform, or radical). In case of mismatch, the project manager has to inform the project owner and project sponsor that the project is starting with undue risks.

Market Service Description (MSD)

Once the decision to carry out the project has been made, the main business assumptions of the project are captured in a market service description (MSD). The MSD describes how the project fits into the product portfolio of the company and describes the target market (in terms of segment and geography) and the value of the delivered service to potential customers. It characterizes the project boundaries in terms of the regulations that need to be followed, the competitive landscape, and the effects of the new service on existing customers or legacy services as well as on the internal operation and support systems. The MSD usually contains a rough order of magnitude analysis (ROMA) of the expected revenues, the financial costs, the resources required to support the service, a risk assessment, a risk reduction plan, and an exit strategy. It outlines the main known risks and the corresponding risk strategies for market entry as well as for market exit. In some cases, particularly for temporary networks, the MSD will describe how to retire the service and the related transition and migration plans.

The amount of details in the MSD depends on the consequence of mid-course corrections on the project success. Plans can be extremely fluid if flexibility is important because of instability in the market conditions or incompleteness in customer requirements. Work can proceed, provided that a change control policy is in place and there is an agreement as to when the basic assumptions can be reconsidered and how to modify the scope of the project. The definition of these criteria is particularly important when the service plans are the result of long negotiations among the stakeholders so as to avoid resistance to future modifications that may be warranted.

Other factors that affect the content of the MSD are the category of the innovation that the project introduces and the overall culture of the project team. When there is a discontinuity in the value chain, especially for architectural innovations, marketing insights are indispensable to reveal hidden needs that new combinations of existing technologies can meet. With respect to the cultural dimension, tight specifications are the norm in the so-called low-context cultures, where the information is coded in words [Hall and Hall, 1990]. In high-context cultures, however, insisting on additional information may be regarded as a lack of trust. This initial lack of precision calls for a rigorous change control policy.

SCOPE DEFINITION

Scope definition is the phase where the project content is defined in the form of a work breakdown structure (WBS) and a technical plan. Let us discuss each item separately.

Work Breakdown Structure

According to the Project Management Body of Knowledge (PMBOK®—ANSI/PMI 99-001-2000), the WBS is "a deliverable-oriented grouping of project elements which organizes and defines the total scope of the project. Each descending level represents an increasingly detailed definition of a project component" [PMI, 2000]. In other words, the WBS is a hierarchical representation of the work contents starting from a very general view to progressively smaller elements, with the elements at the bottom called work packages (WPs).

A work package is typically defined as the smallest work element of a project that can be meaningfully managed, such as the amount of work that a single person can perform in 2–3 weeks. The activity to implement a work package should represent a finite amount of work; that is, it should have clear beginning and ending dates. It should be logically related to its predecessors and successor tasks. To avoid being encumbered by unessential details, the details of the execution should not be included in a work package.

Raz and Globerson [1998] have provided the following checklist to guide the decomposition process. The rule is that the larger the number of positive answers to the following questions, the stronger the justification for further subdivision of the work package:

1. Is there a need to improve the accuracy of the cost and duration estimates?
2. Is there more than one person assigned to the work package?
3. Does the work package include more than one functional activity?
4. Is there a need to know precisely the timing of activities within the work package?
5. Is there a need to cost-out activities internal to the work package?
6. Are there dependencies between the internal activities of a work package and other work packages?
7. Are there significant interruptions in the execution of the internal activities of the work package?
8. Are there different precursor activities to the individual activities internal to the work package?
9. Is there any acceptance testing applicable to the deliverable before the entire work package can be done?
10. Are there any intermediate deliverables that are useful to the project, such as generating a positive cash flow?
11. Are there any specific risks to internal activities that require focused attention?

With the WBS specified, it is possible to propose task durations and outline their dependencies so that the functional teams can give their comments. With the WBS in place, teams can manage the details of the work packages themselves, provided that they use the communication channels to report difficulties and results and request changes when appropriate.

In a typical telecommunications service project, only the highest levels of the WBS are specified when the scope of the project is defined; the details of the work packages are left to the functional organizations that are executing the tasks. Although the rolling wave method (or phase-gate approach) explicitly admits that this decomposition is not available, the implications are not made explicit to the project sponsors and/or customers. In particular, there are two important implications:

1. Because the high-level project planning team does not have enough knowledge to decompose the WBS into its various WPs, the scope definition has a large element of risk, particularly when a new technology or a new market is envisioned.
2. For new designs, details are not defined with precision a priori—that is, before the design has been implemented. In these cases, estimates for cost and resources needed are very tentative and there is a high probability of misjudging the time and cost needed. The margin of error in the ROMA estimates may exceed 100%.

This shows the importance of having the structure to manage the inevitable changes that will be introduced in the scope of the project.

Technical Plan

The MSD triggers the definition of a statement of work (SOW) to specify all the work packages that must be executed and integrated with each other to deliver the planned services. As previously indicated, a telecommunication service includes several layers that need to be defined: the network technology, the operations support systems, and the methods and procedures. Therefore, in addition to the network elements, the SOW covers the network elements management system, the operation support systems (OSS), the training, and so on. Although it is preferred that the service definition be independent of the vendor platform, for many emerging services this is not the case and the definition may vary with the vendor platforms used.

Based on the assumptions and the business objectives, a technical service description (TSD) gives the technical features of the service and its performance expectations in terms of capacities, performance, availability, and reliability. This description is translated into requirements on equipment and/or vendors and drives the development of a series of engineering rules for the network. These rules define the equipment configuration, the quality objectives and quality measures (who records what, how, and where are the records stored, etc.), billing and invoicing, and so on. Of course, the engineering rules will be modified and made more precise as the project team gains experience with the equipment precise during various tests before equipment is certified for the network.

Some of the aspects that the technical description defines are:

- Access to the core network (wireless, wired, high speed, etc.)
- Network topology—for example, redundancy through multiple routes, dual homing to avoid single point of failures, and so on.
- Numbering or addressing plan (if needed)
- Frequency plan (radio and mobile systems)

With respect to the physical installation of the network, the plans describe

- The layouts for the physical installations
- Procurement, warehousing, deployment of the equipment and policies for spare storage and returns of defective cards from the field
- Backup and emergency procedures (power supplies, etc.)
- Training of personnel (help desk, maintenance, etc.)
- Customer support (call center, messaging, etc.)
- Network support (trouble shooting, scheduled maintenance, etc.)
- Change control procedures
- Disaster recovery

Plans for the operation support systems (OSS) cover the following items:

- Provisioning
- Configuration and inventory management
- Network element management
- Maintenance
- Accounting and billing
- Security

THE NEED FOR SCOPE MANAGEMENT

Scope changes refer to variations in the project boundaries due to the addition, modification, or deletion of a work item. These are distinct changes to the project baseline such as milestones or schedules in response to the project environment and [Knutson and Bitz, 1991, pp. 86–108]. The following two examples show the need to take into account changes brought about by industry instability. The first is about the incremental changes that took place during the Salt Lake City Winter Olympics; the second relates an architectural innovation in the area of automatic toll collection on highways.

Salt Lake City Winter Olympics

AT&T was the original supplier of equipment and long-distance telecommunication services for the Salt Lake City Winter Olympics. Unexpectedly, AT&T split into two entities, one for long-distance communications and the second, Lucent, for equipment. Lucent next spun off Avaya to deal with customer premise equipment, while it chose to concentrate on network equipment. As a consequence of all these subdivisions, the project manager for the telecommunications aspect of the Games had to deal with three suppliers instead of one. This meant some changes in the procurement management procedures.

E-ZPass Toll Collection System

Background. In the late 1980s, several approaches were pursued to capitalize on advances in microelectronics, informatics, and telecommunications so as to smooth automobile traffic. One line of thinking was to use real-time traffic information to control the traffic flow, depending on road conditions. Another was to use electronic toll collection (ETC) systems to enable vehicles equipped with radio-frequency transponders—or "tags"—to pay automatically during their passage through the toll station. The transponder would communicate with a reading device (called reader or antenna) in the frequency band around 910 MHz and inform a central computer of its unique ID, while the reader would append a time stamp and a location identifier. For tolls that are distance-based, the computation would use the points of entry and exit. Furthermore, in-lane sensor devices would provide clues to classify the class of each passing vehicle so that the computer could adjust the charge accordingly. Finally, a photographic system would record the license plates of vehicles that pass without a valid tag. The expected benefits were [Ramasamy and Radwan, 2003; Sussman, 2003]:

- An increase in toll-lane throughput from around 500 vehicles/hour to 1800 vehicles/hour, thereby obviating capacity expansions at toll plazas and saving the associated capital and construction cost
- A reduction in motorist waiting time
- A reduction in pollution and increase fuel savings
- A reduction in the cost of toll collections by having less attendants and avoiding of secure cash collection and transportation

The Intelligent Transportation Society for America was formed as a nonprofit organization in 1991 to promote the development of intelligent transportation systems (ITS) in the United States. A federal strategic plan was developed in 1992 to serve as a blueprint for ITS deployment; this plan also influenced the 1998 Transport Equity Act for the 21st Century (TEA-21).

Participants in the scheme would have to subscribe by giving data on their bank accounts to the toll operator and receive in return the transponder. The amount that each subscriber would pre-pay would be stored in a subaccount of the bank account of the toll operator. Should the amount in that subaccount decrease below a certain threshold (around $10), the subscriber's current account or credit account would be automatically debited to replenish the subaccount. Because payment authorization would be made at subscription time, cryptographic protection of each transaction at the toll stations would not be necessary. In addition, the operator would gain an interest on the amounts deposited, thus defraying some of the operational costs.

Gaps in the Definition ITS Scope. From this description, we see that scope of the ITS as defined suffered from three major gaps: the lack of standards, no consideration for control, and the absence of customer care centers.

Lack of Standards. Clearly, ETC is an architecture innovation and, as explained in Chapter 2, there is a tendency to forget standards for these types of innovations. However, the incompatibilities of ETCs on a regional basis cause some inconveniences to commuters as well as to truckers. At a minimum, the transponders should be able to communicate with a variety of readers. Standards would improve the back-office functions (i.e., the OSS) of operators by facilitating the exchange of billing records and account settlement and even enable the integration of common operations to increase speed and reduce the chances of error.

Standards, of course, take time to be developed, tested, and implemented so there might have been some pressure to get a system quickly running and accept a mediocre performance that is "good enough." Unfortunately, in a networked environment, retro-fitting systems for compatibility is an expensive proposition, a situation that Paul Crouch, an ex-colleague at AT&T Bell Laboratories, described as follows: "If you want it bad, you will get it bad."

Lack of Quality Control. There were no specifications on the acceptable performance levels either for the system as a whole or for individual components such as the acceptable ranges for the transponders' reliability, the tolerances' detection accuracy, safeguards for records protection, and so on.

No Customer Care. Customer care is an essential component of telecommunication services offered to the public. Yet, the customer care function was overlooked, perhaps because the customer, in this case the highway authorities, defined the system's requirements without poling the private and commercial drivers. For example, transponders may be of two kinds: active (i.e., battery-operated) or passive. Active transponders have a longer range of communication, but the operational life of their batteries is around 7 years. As the battery weakens, the transponder transmission power decreases and the exchanges with the antennas become erratic. Yet, there was no provision for establishing service points to test the transponders or to exchange them, so the only way that end-users could discover that their tag was malfunctioning would be to receive multiple violation notices that they would then contest!

Scope Creep in New Jersey. The State of New Jersey compounded the above set of problems with the way it handled the contract to build and operate that system. In March 1998, it awarded a contract to MFS Network Technologies for the installation of the equipment and the collection of toll violations, which was then a subsidiary of WorldCom (MCI and MFS Network Technologies had merged in 1996 to form WorldCom). A few months after the contract was awarded, WorldCom sold its subsidiary to Able Telecom Holding Corporation and MFS was renamed as Adesta Communications. After installing 320 out of the 680 toll lanes, Adesta declared bankruptcy in October 2001. The contract was then given to WorldCom because the latter had guaranteed the performance bonds ensuring completion of the work [Larini, 2001]. In the summer of 2002, WorldCom declared bankruptcy and so the project was transferred to ACS State and Local Solutions, a provider of business process and technology outsourcing solutions, in March 2003. Also, to reduce the deficit in the system, a $1 monthly fee was introduced and the discounts to E-ZPass users was canceled [Malinconico, 2002]. Finally, the plan was revised to reduce the number of toll lanes to be automated.

In addition to the instability in the contractor, the basic assumption was that the system would pay for itself through payment violations, so there were no incentives to control the quality of that part of the program. As a consequence, 71% of the violation notifications for the E-ZPass systems in New Jersey were due to the system's malfunction [New Jersey Department of Transportation, 2002]. To offset project costs, a fiber contract was added to be deployed alongside the roads to be leased to telecommunication companies, a good example of scope creep. This fiber capacity was later leased to Rutgers University and the State of New Jersey for their internal communications needs, but the projected revenues could not be realized after the collapse of the telecommunications industry. The optimistic scenario encouraged the State to establish an aggressive payment schedule but as the projected revenues never materialized, payments on the principal were not made and the amount of interest jumped. From 1999 to July 2002, the State of New Jersey had collected $1.7 million in tolls and $14 million in fines but paid $33 million! In reality, instead of paying for itself and generating $34 million of profit, the deficit amounted to $469 million [New Jersey Department of Transportation, 2002, p. 8].

Subscription management and violation payments were outsourced to two different vendors; the J.P. Morgan Chase Bank (now BankOne) and WorldCom, respectively. Without a standardized protocol, communications between the two entities were, to say the least, problematic.

Scope creep continued with the proposal to increase the subscriber's base as a solution to the deficit problem by employing E-ZPass to pay parking fees at airports and sport stadiums.

The main point from this discussion was to highlight the vulnerability of projects lacking a strict and high-quality change control policy. The E-ZPass case provides additional lessons on the way incorrect and unchallenged assumptions threaten the economic viability of a project. It also demonstrates how diffused responsibility can hinder projects governance.

SOURCES OF SCOPE CHANGE

Changes to the project scope are induced by internal or external factors. Internal influences tend to reshuffle the project priorities as a consequence of (a) major reorganizations, (b) changes in sponsorship or in funding, and (c) changes in resources availability that could affect the project parameters in terms of cost, schedule, or the desired quality. More surreptitiously, the excitement of the new project could encourage the bypass of normal organizational procedures and safeguards leading to premature definition of scope [Royer, 2003a]. The experience accumulated during the course of the project may also uncover missing items, clarify incomplete details, or give a better appreciation of the anticipated risks.

External influences include new market demands, changes in regulations through legislation or court decisions, new industry configurations through the alliances of competitors (or their mergers or spin-offs) or their bankruptcy, and so on. For example, changes in rules and regulations may mandate the restructuring of the service offer or may make the project obsolete. Large projects that take several years are susceptible to technology shifts. In the case of infrastructure projects, unusual conditions—whether natural or man-made (such as fires, floods, earthquakes, or extreme weather conditions)—usually perturb the project schedule. Other extreme risks include wars, riots, and so on.

Customer Profile

Scope creep typically results from changes in the customer profile or expectations. Several factors contribute to an incorrect understanding of the customer's requirements [Frame, 2001]:

1. Focus on the wrong applications or lack of knowledge of the end-user's applications.
2. Addressing the wrong set of customers or end-users, particularly if the client representatives do not fully understand or adequately represent their needs.
3. Contradictory requirements because each set of users and customers have distinct requirements
4. Requirement instability because the initial description of the needs or because was incomplete.

Disruptive innovations, whether architectural or radical, are particularly susceptible to wrong market assumptions. Rapid prototyping methodology can be used to test the response to architectural innovations. For sustaining innovations, techniques such as Quality Functional Deployment (QFD) or benchmarking competitors can be valuable in defining the features to be included in the new service.

Vendor's Effect

Changes to the equipment because of the adoption of a new standard or the introduction of a new technology could force a redesign and a review of the project scope. Implementations difficulties that would prevent the vendor from supplying a feature on time could be another factor. Mergers and alliances among suppliers could introduce changes to the set of assumptions of the project, including equipment discontinuity.

The diagram of Figure 4.1 summarizes all these changes.

Basic Principles of Scope Management

To handle all the factors, a strong change control policy is needed. The guiding principles for managing scope change are as follows:

- Subjectivity is an inherent component of the decision to launch a project, even though rational criteria provide some foundation. The subjective assessment relies on experience, understanding of trends in technology, individual ambitions, and ideological or sociopolitical motives such as to outsmart the competition, maintain employment, gain markets, and so on. Subjectivity, however, should be less significant in the case of incremental innovations, provided that good historical records are available.

- Because of this subjectivity, a process is needed to make sense of incoming information and modify the project scope accordingly or—in extreme cases—end it.

- Project estimates in terms of cost and duration should be considered as living documents to be updated. The frequency of updates depends on the type of the innovation and the level of the WBS at which the change takes place.

- The cost of modifications to the project plan increases with progress in the project implementation. A tracking and monitoring system should serve as an early warning system.

- Once problems have been anticipated or detected, resolutions must be sought to rectify project trajectory. These corrective actions, however, must go through sufficient analysis and discussions before any agreement and an audit trail of all discussions and decisions must be kept for the future. Without such a discipline, reactions to pressure without the benefit of wide deliberations would confuse or demoralize the project team or introduce other problems.

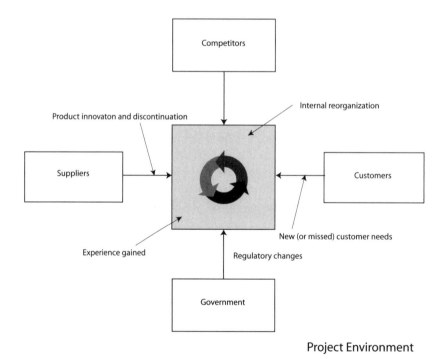

Figure 4.1   Factors of change in the project scope.

We now review the source of changes that can take place during the implementation of the project and their impact on the project scope or baselined plan. Then we explain the steps to ensure the constancy of purpose through a change control procedure.

Change Control Policy

Change requests are made when the project has irremediably deviated from its baseline. Change control—also called *configuration management*—is a documented set of procedures that define how to treat these changes. In some cases, configuration management is reserved for the functional and physical characteristics of the deliverable [Rad, 2002, p. 83]. Irrespective of the terminology, the change control policy should cover the following items:

1. The list of stakeholders that can request a change. Formal change requests are called by different names such as engineering change order, modification request, and so on. In typical telecommunication projects, engineering change orders refer to hardware changes, while modification requests are used for software.
2. The ways to document a change request including its urgency and severity, the reasons for the request, and its potential effect on the project cost or schedule. Many software programs assist in the design of electronic change control forms, routing them to their intended recipients, tracking them as discussions take place and decisions are made, and then finally archiving them.
3. The process for examining and validating the request. The policy should specify who validates the requested changes, the intervals for decision, and the approval authorities.
4. The method for documenting the decision and its consequence. This is essential to ensure continuity because membership in the project team changes over time.
5. The communication process to announce the proposed changes, solicit comments, and distribute the decision and the approved, if any, changes to the various stakeholders, including the customer and/or project sponsors. Software tools can manage the modification requests from their arrival to their resolution and keep an audit trail to link the decision to the modified baseline documents for traceability.

Strictness of the Change Control Policy. Without a change control policy—that is, a disciplined approach to process change requests—confusion and demoralization will befall the project team. At any level of management, the need for making the change should be articulated and its impact on the project's cost and schedule evaluated. This allows all the stakeholders at that level the chance to evaluate the request before making a decision on whether to accept, modify, or reject the request for change. Any decision made should be documented and communicated to all interested stakeholders.

While all projects need some kind of change control, the strictness of the policy is more important in the following instances:

1. In government projects, all expenses need to be justified.
2. The roll-out of the project takes several years because no matter how much thought went into the original plan, changes are unavoidable.

3. The initial plans, as outlined in the MSD, were fluid to ensure flexibility and to start development as quickly as possible.

4. Market conditions were not known or the customer's requirements were fuzzy.

5. Because of any set of considerations, many assumptions were implicit and not recorded in the MSD.

On the low end, a project may have no project charter and the scope of its activities may be defined ad hoc. On the high end, a model project would have a project charter, a scope definition, a baseline plan, procedures for evaluating changes and distributing the relevant decisions, and a well-defined matrix of roles and responsibilities. In this way, as plans get updated, members of the project team will be able to have an up-to-date version at all times.

Change Control Board. One way to streamline the process of change management is to centralize the responsibilities for managing changes in a change control board. This board is formed from representatives of various organizations involved in the project execution (service development, system engineering, field support, testing, network management, etc.) as well as the project owner or sponsor. The board convenes on a regular basis to evaluate the change requests and make decisions regarding all pending change requests and to review the status or progress of all outstanding change requests. This allows stakeholders' participation as an equal partner in all meetings, reviews problems, issue resolutions, and decisions so that they all share the responsibilities for the product.

Once a change has been approved, it is communicated to the various stakeholders as indicated in the communication plan.

SCOPE VERIFICATION

Scope verification is done throughout the project execution by comparing the planned project scope to the actual project results before deciding whether to accept the work performed at key milestones and at the end of the project.

Tracking and Issue Management

As the project progresses, some issues inevitably arise that need attention over some span of time—for example, following a purchase order, the delivery of an equipment, and so on. Data collected during the course of the project can alert the project manager to critical issues or impending difficulties, provided that they depict a realistic view for the progress in terms of budgets versus actual costs, planned work versus actual accomplishment in terms of starting dates and ending dates, amount of work, resource utilization, and so on. These data would provide a historical base for streamlining and managing of future projects.

Tracking and monitoring consists of two main items: data collection and processing and then information distribution to team members and stakeholders, to keep them aware of the project constraints and current view regarding the key milestones, target costs, and the feasibility of the technical requirements. The role of the project manager is to integrate the various streams of subactivities into a grand scheme of things so as to evaluate the project status on a periodic basis and make appropriate decisions accordingly. All these are time-consuming efforts, even though on-line tools facilitate the task of data acquisition and distribution.

The progress monitoring system should be designed and implemented to facilitate the work and not impede it. In other words, it should not affect the team's morale and enthusiasm by appearing as an unnecessary burden, an impediment to creativity and innovation, and a way to snoop on them because of the lack of trust. Thus, the introduction of such a system should be gradual and flexible and should avoid using the data collected for individual performance evaluation or to pressure for increasing the productivity without resolving the systemic root causes. Negative attitudes of the team to the progress monitoring system can lead them not to provide the data regularly and even not using the system or providing meaningless results just to improve the numbers.

While the detailed design of such a system will depend on the cultural environment in which the project takes place, it must ensure the following functions:

1. Inform team members of individual assignments.
2. Remind them of forthcoming events.
3. Track and document issues raised and their resolutions.
4. Collect performance data and feedback from the data.
5. Process the data and distribution the information on variances and any changes to the plan.

Issue tracking at any level of responsibility requires the definition of (a) the authorities that can open or create an issue, (b) how to define the urgency or the severity of an issue, and (c) who can assign the issue to an individual for coordinating the resolution. When an issue is closed, documentation of the resolution should be included. Depending on the severity of the issue, higher levels of authority may be required to sign off on the resolution. Performance of the issue resolution process is evaluated in terms of the total number of issues created, the number per severity class, how many issues remain open, how long it takes to resolve an issue, and so on.

Project Termination

Changes in the project scope have implications on the management of a project, some more serious than others. They can affect the service definition such as the set of features offered or the pricing. This shows that change management and risk management are related. Dealing with risks imposes changes, which, in turn, could introduce additional risks. Clearly, risk management and change management take place continuously during the life of a project. It should be kept in mind, however, that the later in the project cycle that changes are made, the costlier they would be.

Projects may have to be terminated for reasons that have nothing to do with the project performance, because the assumptions that were the foundation of the project charter are no longer valid such as:

- Changes in the technology may have rendered the planned service obsolete.
- Unexpected technical or technological difficulties have made the service uneconomical.

- The market expectations and customer profiles have changed.
- The project sponsor may have lost interest or power or moved to a different company.
- The company structure has changed through an alliance or a merger, and the new leadership team is no longer interested in the project.

CASE STUDIES

In a survey of 70 professional engineers conducted in 1994, the No. 1 rated reason for project failure was that "the project was not adequately defined at the beginning." The third most highly rated reason was "a lack of clearly defined project goals and objectives." The fifth reason "project planning was done with insufficient data." All these statements related to the scope of the project [Black, 1996]. While it is unlikely that the project planners would have detailed knowledge of all the factors needed at the time the project is defined, the real thrust should be to establish the necessary structure to manage the necessary changes and steer the project in the right direction.

Experience is the best teacher, and business school cases are usually devoted to success stories. However, just like the study of diseases gives a better understanding of the human body, project failures can be illuminating. One reason is to avoid repeating the mistakes of the past; another is that mistakes of the past have fashioned the present. This is why we present in some detail two cases to illustrate the importance of good scope definition and management through controlled changes.

Telecommunications Alliances/Joint Ventures

The general precepts for alliance-making are: (1) Be specific about what is wanted from the partners, (2) expand the scope of the alliance gradually after an early success in a narrow aspect, and (3) go for the market leaders, not the followers [Wheatley, 1999, p. 125]. However, relaxation of the regulatory environment in the 1980s and early 1990s opened so many new opportunities beyond the capabilities of any single operator that many telecommunications companies chose to ignore past wisdom and rushed to form alliances and joint ventures to offer global connectivity services to large multinational corporations [Cowley and Aronson, 1993, pp. 190–214]. As shown in Table 4.1, a large number of these attempts fell short of meeting their corporate objectives. This less than stellar record should be compared to relative successes of collaborative efforts among equipment manufacturers. For example, Bluetooth was a joint effort of many manufacturers led by Ericsson, Nokia, and Intel to connect mobile phones, computers, and peripherals without wires [Keil, 2002].

BT started a number of joint ventures mostly in Europe and established Synchordia in Atlanta to provide global outsourcing services and bought Tymnet. AT&T created the JENS network in Japan in partnership with Japanese trading houses and purchased ISTEL in the United Kingdom to link the two regional networks through the United States. It then tried to build its own network in the United Kingdom through AT&T-UK, then started an alliance with Unisource and WorldPartners before switching to a joint venture with BT (Concert) and then retrenching back to the United States. The failure of all these alliances has been considered as indicative of the pervasive uncertainty and instability on

Table 4.1 Failed Global Alliances (1985–2001)

Alliance Name/Parties in the Alliance	Start Date	End Date
BT and EDS		1993
Telecom Italia and IBM		1995
BT and Portugal Telecom		1998
Telecom Italia and Cable & Wireless		1998
AT&T, BT, and KDD (GIMM)	1985	1998
Unisource (KPN, Telia, Swisscom)	1991	1999
Atlas Communications (Deutsche Telekom, France Telecom)	1993	1995
WorldPartners (AT&T, KDD, Singapore Telecom)	1993	1998
Concert (BT, MCI, PT Austria, Portugal Telecom, Tele Denmark, Telefónica, Telenor)	1994	1997
Global One (Deutsche Telekom, France Telecom, Sprint)	1996	1999
Telefónica and Unisource	1996	1997
Telecom Italia and AT&T	1996	1998
AT&T and Unisource	1996	1998
Telefónica and Concert	1997	1997
Concert (AT&T and BT)	1998	2001
Vizzavi (Vodaphone and Vivendi Universal)	2000	2002

Sources: Curwen [1999], Probe Research [2000], and Jolly [2002].

the future shape of the telecommunications industry [Curwen, 1999]. It was mentioned that the objectives of the alliances were not clear; there were latent competitions in the alliances, and the relationships with the parent companies were problematic. This is an indication that the scopes of these ventures were not adequately defined.

Another major hurdle that these alliances encountered was the inability to pool resources and achieve economies of scale, even though all of them shared the same characteristics: mostly incumbent telephone companies, long experience, steady revenue stream, and so on. However, each company had developed its own methods and procedures (M&Ps) to streamline the hundreds and thousands of processes that support business delivery. These processes had to be reevaluated and merged across all parties. Furthermore, when new technologies are introduced, considerable attention has to be devoted to redefinition and enhancement of business processes.

Even if we overlook the issue that each partner may have a different strategy, merging services and networks are very complex operations that require careful and time-consuming planning. For example, the joint network may not have the same capacity to fit the traffic needed, so this means that for a very long time, several networks will have to coexist and work seamlessly with each other: one from each partner and a joint network. This increases the attention to be given to the operations and maintenance of these networks without necessarily improving the quality of the services offered.

In the case of incumbent operators, compiling the details of existing network is a time-consuming task. For example, the number of billing and operation support systems (OSS) platforms is typically very large (about 800 for AT&T [Nairn, 2004]). Billing systems are particularly complex because there are many tariffs for different customer groups and their profiles changes with industry trends. Negotiations with each business customer are needed to plan the transition from one billing system to anther. Actions need to be tightly coordinated so that the old billing system is disconnected exactly at the same time that the new network is established. Billing reviews are needed for every account to ensure that the old network is indeed disconnected and that the systems of the new network are operational. Resources need to be in place to handle billing disputes. Hamid Akhavan, CTO of T-Mobile, explains that merging different billing and OSS is "a very long route and it can take several years" [Akhavan, 2004]. The reason for this long migration path is that a common process has to be agreed upon and then moving to that system will require operating both old and the new system for a while until all the problems have been ironed out and all people have been retrained. In fact, when Cingular acquired AT&T Wireless in 2004, it was largely anticipated that the cost of integrating the two networks and their support systems would outweigh any savings from the merger for at least two years [Taylor, 2004].

From Table 4.1, it is clear that very few telecommunication alliances lasted long enough to deliver their full values. The rapidity of the successive changes betrays a short-term view and a fecklessness that are not compatible with the nature of telecommunications operations. Simply put, those who planned these alliances did not seem to have considered what it takes to integrate the service delivery processes of the alliance partners. This explanation seems to be supported by what happened in the late 1990s when AT&T attempted to build a broadband local access through a merger with TCI without adequate planning; in this case, technical considerations were brought about by the business press and industry analysts that questioned the costs of these ambitious plans [Martin, 2005, pp. 123, 133–136]. Clearly, "most business collapses result from poor judgments. The company mismanages the nuts and bolts of corporate organization—financial controls, personnel management or customer relationships" [Kay, 2004].

Alliances in wireless telecommunications offer an interesting contrast in that they have a higher success rate, sometimes ending in full mergers. The reasons could be that mobile operators had no legacy systems, were facing few legal constraints, and were obliged to cooperate to extend their zones of network coverage. It is also possible that the way the mobile companies were managed as separate entities from their parent companies or from fixed-line operators allowed them more flexibility in responding to the problems as they arose.

Net 1000

We use this case study to emphasize the importance of scope definition and change control. The following information is based on public information [Ricci, 1984; Dix, 1985; Howe, 1986] supplemented by a postmortem analysis presented to Ian Ross, then president of AT&T Bell Laboratories [Wilcox, 1987].

The yearning for an all-digital intelligent network to tap distributed knowledge spread over disparate machines and programs was a topic of active research since the late 1960s. Many networks such as BITNET, DECnet, and so on, were adopted to handle business exchanges within a company while the System Network Architecture (SNA) defined the protocol suite used to connect IBM mainframes.

AT&T tried to enter the fray with Net 1000, a commercial data networking services available from 1982 to 1986. The net spending on the project was evaluated at $450M, while the overall expenditure over a period of 10 years was about $1B.

Net 1000 was not the only commercial failure on the road to digital connectivity. An end-to-end digital network on microwave links, Data Transmission Co. (Datran), ran out of money in 1976 after eight years of operation. Xerox's network for digital document exchange with microwave links was abruptly terminated in 1981 after merely two years. The U.S. Postal Service introduced a hybrid electronic mail letter carrier service ECOM that did not last long [Howe, 1986]. The Minitel was a partial success in that it remained confined to France and could not be exported. Unfortunately, the project management literature tends to overlook the valuable lessons learned from unhappy experiments so that the same mistakes are repeated over and over. In the decade after the demise of NET 1000, most studies were conducted on the way to obtain digital information infrastructure particularly, with the decision to commercialize the Internet. This includes the Information Infrastructure Project at Harvard University or the studies associated with the National Information Infrastructure (NII) initiative that the Clinton–Gore administration launched in September 1993. With the extreme focus on attracting national attention to the need for a digital infrastructure, the valuable lessons from NET 1000 were lost and many mistakes were repeated on a grand scale in the dot com boom and bust.

Background. In 1934, the U.S. Congress established the Federal Communications Commission (FCC) and defined a national goal of making basic telephone services affordable anywhere in the country. This goal was called *Universal telephone service*. To achieve this, AT&T was given the monopoly of telephony in most areas; however, this monopoly was under challenge by the U.S. Department of Justice using the anti-trust provisions.

In 1956, AT&T signed a Consent Decree to settle an anti-trust suit that the U.S. Department of Justice had filed in 1949. The agreement allowed AT&T to retain its manufacturing arm (Western Electric) as long as it would sell its products only to the Bell operating companies. Furthermore, the Bell System was confined to common carrier activities and barred from providing commercial data processing services.

Outside the Bell System, value-added networks (VANs) filled the data processing needs in the banking and financial sectors, as well as in the airline and automobile industries. Initially, VAN leased circuits from the Bell system on which they overlaid their offers. The next wave of VANs allowed the sharing of applications among many users on a commercial basis (e.g., the SABRE reservation system of American Airlines). Time-sharing service companies set up in the 1960s made the usage of powerful host computers affordable through economies of scale by sharing the hardware and the software among many users. Finally, Electronic data interface (EDI) emerged to streamline business communications through electronic documents as they related to ordering, shipping, receiving, billing, and payment. It was observed that transaction automations reduced cost and increased processing speed by avoiding repeated reentries of the same data, validating the data elements according to common rules, and automatically routing them to the destination.

Progress in computer communications blurred the differences between the activities that were forbidden to the Bell System (data processing) and those that were allowed (data communication). At the same time, computer manufacturers were able to off-load the communication handling tasks in large hosts to specialized processors called communication processors. With expanding customers needs, the convergence of communications and data processing led to the development of the concept of a "shared intelligent" data network. While telephone penetration in the United States reached more 80% of the households by the mid-1960s, the Bell System chaffed under the restrictions that prevented it from providing new services that needed processing of user voice traffic within the network. For example, voice messaging or voice response systems (where tones could be used for order entry with guidance from stored voice prompts) were not possible within the existing regulatory regime. The thrust for changing the public policy concerning common carriers came from many directions.

New networking concepts came out of the work that began in the late 1960s and continued through the 1970s under the sponsorship of the Advanced Research Projects Agency (ARPA) of the U.S. Department of Defense. An experimental network, the ARPAnet, using packet switching, could avoid the underutilization of networking resources typical of traditional circuit switching techniques when used to transport data, thereby making it affordable to a wide variety of commercial application. Electronic mail was invented in 1972. At about the same time, Ethernet local area networks (LANs) were developed to connect desktop workstations in a given area together. All LANs within a building or a campus were then connected to the wide-area network. With the availability of high-speed transmission on fiber-optic cables, the National Science Foundation (NSF) funded a Computer and Science Network (CSNET) that evolved into the NSFNet in 1986 to allow research centers to share scarce resources such as supercomputers. The combined impact of all these activities was to show that the capital and operational expenses associated with advanced computing systems could be shared among many users. They also showed the need to update and modernize the regulatory policy.

In an attempt to resolve the regulatory aspects, the FCC began the first of three computer inquiries to explore how to change the regulations of the Bell System and avoid choking innovative telecommunication services [Rey, 1983, pp. 702–704].

In Computer Inquiry-I (1968–1971), the FCC distinguished between telecommunication services and data processing services. The first set could be provided by the regulated telecommunications industry. Companies providing data processing services would remain unregulated. Services judged to be hybrids would be determined on a case-by-case basis.

Having found itself swamped by requests to evaluate these hybrid cases, the FCC initiated a second activity, called Computer Inquiry-II (1976–1980). During that time frame, another anti-trust suit was filed against AT&T in 1974. Following this inquiry, the FCC imposed a structural separation between basic services and enhanced services. Basic services were defined as those where the network behaved as a transparent pipe providing simple transmission without modification of the content. Enhanced services were defined as those where the network acted on the content or format of the transmitted information to provide protocol conversion, information storage, or packet switching. Regulations would apply to basic services only. In addition, the Bell System could offer enhanced services or sell customer premises equipment, including ordinary residence telephone, through a "fully separated" subsidiary. This subsidiary was to be operationally self-sufficient so that the regulated part of the business could not subsidize the enhanced services. Thus, this entity would have to have independent organizations for sales, product management, research and development, network operations, and administrative support. NET 1000 was the Bell System's—and later AT&T's—attempt to take advantage of the relaxation in regulation to enter the area of data processing.

To settle the 1974 anti-trust suit, AT&T agreed to divest itself from its local telephone companies starting January 1, 1984. Finally, after the breakup of the Bell System, Computer Inquiry-III (1984–1986) eliminated the subsidiary requirement and substituted special accounting rules to prevent cross-subsidies of nonregulated services.

Timeline and Organization Evolution. On July 1, 1982, American Bell was formed with about 1000 people as a fully separate subsidiary of AT&T to offer Net 1000 and any other future enhanced network services within the guidelines of Computer Inquiry-II (CI-II). Its only working relationship with the rest of AT&T was as a tariffed customer for basic network services and as a large customer for end-user terminals (modems, phones, etc.).

The service offer consisted in providing customers with the capability to develop, install, and manage application software to run on AT&T's owned processors. The architecture was based on having a large number (100–200) of disperse data centers (called "service points"). These were interconnected using an X.25 packet switched network from the regulated part of AT&T. Initially, data centers were built in New York, Chicago, Los Angeles, Greensboro, Salt Lake City, Camden, Kansas City, and San Antonio. Statistical multiplexers in remote access points brought the total number of service points to 45. A Network Operations Center was constructed in Somerset, NJ.

Net 1000 was touted as "the cornerstone" of AT&T's data strategy to sell data processing services through its new unregulated subsidiary and compete head-on with other VANs. During the first four years, it was expected that customer demand would be growing geometrically for the first few years and capacity growth was planned accordingly.

The idea of Net 1000 was for users to pay for what they use. They were charged for network terminations (ports), disk storage, transmission bandwidth, connection time, and communications processing.

Six months later (January 1983), the mission of American Bell expanded to include customer premises equipment such as terminals, modems, private exchanges (PBXs), and so on; senior management was changed to fit the new scope. AT&T Information Systems would then compete, not only with VANs but also with established vendors such as IBM and DEC. The sales organization dedicated to Net 1000 services was integrated into the overall sales force responsible for data networking services and the associated equipment.

The lack of customers in the 1982–1983 timeframe provoked consternation. A retrenchment strategy was executed to resize the project and refocus its activities. Expansion plans were canceled or postponed; unused nodes were closed (Kansas City and San Antonio) and the network operations staff was trimmed. One side effect of this retrenchment was to keep the number of service points at 45 instead of the designed 200.

To address the lack of applications, the offer was reconfigured to be a vertical integration of PCs, software applications, and data networking services. The software architecture was changed to the emerging client/server architectures, with some processing on the customer PC. The server would then do certain store-and-forward function as well as some routing. EDI applications were seen as a way to provide some market pull because once a major company would join, it would pull in its business correspondents and suppliers. EDI offers were tailored to specific industrial sectors (e.g., mortgage industry).

The divestiture of AT&T on January 1, 1984 triggered another internal reorganization with American Bell becoming AT&T Information Systems because AT&T was prevented from using the "Bell" name and symbol. Several months later (May) another reorganization took place.

The marketplace was full of images of the "Clash of the Titans" between AT&T and IBM. For its part, IBM acquired the Rolm Corporation, a maker of enterprise telephone systems, and established a satellite communication company. In the marketplace, there were trials of the service with Ford Motor Co., Dow Jones and Co., Northwest Mortgage, and so on. Ford's objective was to link its auto dealers to locate and exchange parts. Dow Jones, which publishes the *Wall Street Journal,* was aiming at providing customer access to Dow Jones News Retrieval, its computerized financial information service. Northwest

Mortgage, a Minneapolis-based mortgage company, was using it to send mortgage rate quotes and truth-in lending information on loans to its branch offices across the United States. Net 1000 was used also for transaction services (credit card and debit card transactions as well as credit card verification).

The first public setback was a front-page article published on July 13, 1984 in the *Wall Street Journal* detailing the problems that Net 1000 was facing. In October 1984, Ford decided not to use Net 1000 in its dealer's network because "the service was not ubiquitous and it did not have enough nodes" [Dix, 1985].

On January 1, 1985, another internal reorganization took place. In August 1985, a new management team was installed to review Net 1000 prospects, particularly its EDI applications with the help of a McKinsey consultant. The result of the study was to cancel the project and to migrate existing EDI customers to other VANs. Withdrawal of the service was announced on January 20, 1986 with a termination date on May 31 of the same year.

The net spending on the project is evaluated at $450M, corresponding to a loss of $0.6B (post tax loss $250M).

Postmortem Analysis. The problems that Net 1000 faced can be grouped into three categories: business, technical, and regulatory. In the business category, the main problems were:

1. Incorrect assessment of demand
2. Unrealistic revenue expectations
3. Complex sale channels
4. Haste and aggressive deployment
5. Neglect of available expertise
6. Pricing
7. Weak sales organization
8. Unstable organization

On the technical side, the risks due to technology changes were not properly considered—in particular, the shift from mini- to microcomputers and the emergence of the client/server architecture for applications development.

Finally, regulations caused access problems that could not be easily resolved.

We now address these problems individually.

Business Problems

INCORRECT ASSESSMENT OF DEMAND. The computer press was full of images on the "Clash of the Titans," anticipating an epic confrontation between AT&T and IBM. Many were rooting for AT&T to stand up to IBM and take a healthy chunk of the desktop market. The business case was based on customers developing their own Net 1000 applications; so this was a case of an architectural innovation looking for a market pull through identification of needs that need to be satisfied with a new combination of existing technology. However, in disruptive innovations, traditional market surveys may be biased against the innovation. In this case, typical surveys and hundreds of customer interviews missed the market opportunities because the real end-users (the programmers) were not involved and their needs were not considered. As a consequence, there were no training courses on programming Net 1000 applications and the time needed for developers and users to learn on their own was significantly underestimated. Finally, there was no systematic process to track changes in the market assumptions due to developments in the desktop market and no feedback to the scope definition until it was too late.

Ultimately, the view that there was a demand for networked data services was correct, as proven by the popularity of the Internet. However, setting unrealistic goals and objectives at that early stage increased the costs and caused misallocation of resources that ultimately led to the project failure.

UNREALISTIC REVENUE EXPECTATIONS. High margins and fast growth do not usually go hand in hand: High margins are usually for mature services. Nevertheless, the business case projected that revenues of Net 1000 would exceed $1B/year by 1988, even though each customer terminal would generate a monthly average revenue of $100. In making these predictions, the business planners did not consider the absence of application software and overlooked the time needed to develop, test, and deploy software applications, particularly in a new operating environment. The time needed for ramp-up and development of applications was overlooked. To see how these decisions were made, it may be useful to remember the context where with the divestiture around the corner, investors were betting on the competition between AT&T and IBM.

This optimistic view caused haste in the planning and deployment of the network nodes and started fast track (i.e., costly) development programs in addition to a buildup in the sales force and other support functions. Schedules were crushed and resources worked overtime. This caused escalation in expenses without commensurate revenues.

When market reaction was lukewarm (i.e., customers did not stamped within the first 6 months of the network existence), management panicked and took a series of reactions that could not change the fundamentals.

In the case of the World Wide Web, it took several years (4–5 years) for commercial applications to appear. As a benchmark, consider the case of Amazon, one of the most successful Internet businesses. By the end of 2000 (after 5 years of operation), Amazon had sales of $1.6B); its first profitable quarter was the third quarter of 2001, but it did not bring any net yearly profit until the end of 2003.

Given that fact, it is unfortunate that, in many companies, projects have to present unrealistic expectations just to secure funding. However, this may be a pyrrhic victory, because the project could get caught in the vicious cycle of trying to meet or beat expectations, even though this could derail it off the right track.

HASTY AND AGGRESSIVE DEPLOYMENT. AT&T worked feverishly to meet the market expectations. This has led to many unwise decisions that had significant effects on the commercial viability of the service.

In the rush, some unneeded capabilities were developed, while more relevant ones were dropped. Construction of sites went ahead without fully baselined requirements and without adequate risk analysis. For example, a data center in New York City was situated underneath an ice-skating rink and was flooded when the ice-making equipment malfunctioned.

The pressure to compress the schedules caused many fast track activities to be redone to accommodate unforeseen or overlooked requirements. In the heat of the activities, communication of data to members of the projects suffered. For example, there was no repository for data related to sales, current operations, training material, technical information, and so on.

Overcapacity increased the costs in several ways. One of them was unnecessary software development. Also, premature equipment purchases missed advantages due to performance improvement and price reductions that could have been gained just by postponing purchases until real need arose.

Finally, the original name of the service is another indication to that haste: It was announced as "Net One" but was then modified after the discovery that there was a local area network (LAN) product called "Net/One."

NEGLECT OF EXISTING EXPERTISE. Net 1000 was used as a "weapon for changing" AT&T's culture by a team led by Archie J. McGill, a former IBM vice president who jointed AT&T and championed the product. The idea was to teach a "bureaucratic monopoly" how to survive in a world of competition. However, there was some severe underestimation of the complexities of public data services as compared to value-added services in an enterprise network. While both types may share the same set of networking technologies, public services depend on many supplemental systems to manage the operation and administration of the network and meet the quality objectives sold to the customer such as mean time to repair, billing accuracy, and so on.

There was a lot of tension between Bell Labs and the new team. For example, Carl Soutbard, one of the early architects of Net 1000, commented derisively: "Bell Labs thought the Communications Act of 1934 gave them a sacred responsibility to determine what products the country needed" [Ricci, 1984]. The conflict consumed much time and energy and blind sighted each side to the strengths that the other was bringing to the table. One consequence of this dissonance is that the Net 1000 architects overlooked some basic principles of public networks. There were no plans for fault management and remote diagnostics to the point that many data centers (service points) were at unattended remote sites or sites with little or no on-premises maintenance coverage (e.g., sales office). This increased the mean time to repair.

PRICING. The pricing structure required an accounting infrastructure to track usage. One particular problem was caused by inclusion of the processor loads in the equation. Such loads fluctuated according to the application and how the program was executed. As a consequence, customers could not predict their bills, which had to be explained to them. Another problem was that there were no concept of multiple service classes to tailor the pricing to different business needs.

It should be noted that user-based pricing is still a problem in data network, because counting the resources used is usually more costly than sending a packet. A flat rate per transaction could have resolved that difficulty as long as the network was not congested. While this pricing allowed the Internet to be readily accepted, it led to unpredictable network conditions that make it unsuitable for mission-critical applications.

WEAK SALES CHANNELS. One hurdle that Net 1000 faced was that it competed with the lucrative private-line business that was sold by the regulated part of AT&T. Even today, private lines are still the dominant component in data networks. For example, worldwide revenues in 2001 were $25.9B, or 61% of the total data traffic according to the Vertical Systems Group [Dunn, 2002].

Sales support functions were inadequate. These include pre- and post-sale activities to understand user's needs, select the appropriate network configuration, install the software, and train the users. Sales support is essential for business applications such as for

EDI. EDI is not merely a technical arrangement application; a typical sale usually involves multiple companies and multiple departments within each company. It involves political negotiations across organizational boundaries to redefine the workflow and, indirectly, the balance of power. As a result, EDI services require long selling cycles that do not fit the typical sales compensation plans. Not surprisingly, when Net 1000 salespeople were integrated into the AT&T Information Systems general sales force and their portfolio included premise equipment and software in addition to Net 1000 services, their attention shifted to the traditional products that were easier to sell.

The weakness in the sale force meant that there was no systematic way to qualify customer's inquiries and focus on the more promising leads. For example, a deal was conducted with a Tennessee-based hardware distributor (Belknap Tools) to provide an EDI package on Net1000 with a video-quality retrieval system. However, after the development was done, the company went bankrupt and was subsequently liquidated.

The lack of good sales organization and the desperate chase for customers led to a continuous stream of changes in the technical requirements and costly software customization. This added unnecessary costs, particularly because the change control policy was lacking.

UNSTABLE ORGANIZATION. With extremely optimistic assumptions, the initial plans called for doubling the capacity every 6 months irrespective of the market conditions. In hindsight, these objectives were not realistic, but what made the situation worse is that there was no formal policy for reevaluation and readjustment based on what really happens in the field. The only way to modify the course was through replacing the management team, a remedy that brought about a series of disruptive reorganizations (on the average, one every 6 months). With this level of churn, constancy of purpose was not possible because of the significant shifts in focus and in priorities with each new management team.

Technological Risks. The Net 1000 service architecture called for network-based hosting of applications, user's programs, and data as well as for performing the maintenance and data back-up operations. The end-users would customize the interfaces to these applications to suit their environment. To use current terminology, American Bell/AT&T Information Systems was supposed to act both as an application service provider and a data center service provider (ASP/DSP) while the regulated part of AT&T would be a network service provider (NSP) offering pure transport, switching, and store-and-forward functions.

Each service point comprised one or several "nodes," each consisting of five minicomputers (VAX 11/780 and later VAX 8600) with a 4:1 redundancy and their associated peripherals. The operating system was DEC's VMS. There were five or more IBM Series I computers per node to act as the communications interface (front-end processors). COBOL was the application development language.

In many of these technologies, standards (whether formal or de facto) were lacking. For example, protocol conversion was needed to enable asynchronous terminals to access the bisynchronous applications on 3270 IBM computers using the BISYNC protocol. During that time frame, standardization of EDI messages had not sufficiently progressed and the parties of a transaction had to agree among themselves on the syntax of the messages as well as on the rules for electronic exchanges (the responsibilities of each party, the rules of identification and authentication of the various entities, etc.). Substantial efforts were needed to verify the compatibility of the Net 1000 software with the software used at the customers sites.

The Net 1000 architecture was proposed at a time of technological discontinuities, which increased the technological risks. The emergence of microcomputers such as IBM's PC and the client/server architecture challenged the roles of large hosts as well as those of minicomputers. Services that did not take into account the increased computing resources available at the desktop were being challenged. Thus, it was very difficult for anyone, first of all customers, to get a comprehensive view of where the technology was going; this uncertainty distracted customers. One way out could have been to develop the service in a series of incremental capabilities as the market matured. However, the pressure to meet market expectation and deliver a nice product that would awe friends and foes was too strong to overcome. Changing the software architecture to a client/server situation required the restructuring of the application.

Regulatory Constraints on Access. VAN services require an economical access network, and for years the networking community wanted the telephone companies to offer direct data connections that would be as easy to order as a phone line. Unfortunately, FCC regulations prohibited Net 1000 from offering service directly to customer's premises. Furthermore, the regulated side of AT&T could only provide pure transport—that is, no speed or protocol conversions.

To add to the formidable access challenge, due to cost retrenchment, Net 1000 had to contend with 45 service points instead of the required 200–600. This affected customers' perception as to AT&T's commitment to the service. In many cases, the business proposition of Net 1000 was not appealing because of the lack of an extensive geographic presence. Access could be achieved through either private lines or dial-up calls into the nodes themselves. Private lines to service points were economic for large customers. Dial-up customers had to pay toll charges to connect to distant service points.

Plans were in place to use X.25 concentrators to increase geographic coverage when the project was canceled.

Lessons Learned. When Hutchinson 3G launched its third-generation (3G) wireless service, it deliberately used a pricing scheme that did not depend on the amount of data exchanged so that the end-user could work-out in real time how much a particular service would cost. The operator opted for pricing per event; that is, the subscribers would be charged per film clip or per game downloaded. Thus, they have avoided one of the problems that the Net 1000 project struggled with, even if they did not learn directly from this experience.

Lessons Not Learned. Many of the technical problems that confronted Net 1000 were either resolved or on their way to being resolved by the time the NII was launched. The World Wide Web (1991) provided a way to locate resources on different computers connected to the network, and the Mosaic browser (1993) provided a simple, graphical interface to these resources. Many nontechnical users were already familiar with on-line transaction processing while client/server architecture was commonplace. In many areas, standards were in place or being developed: HTML, XML, Application programming interfaces (APIs), and so on.

Despite these technical advances, hype, unrealistic expectations and uncontrolled changes to the specifications had dramatic consequences. Like the case of Net 1000, the dot com mania can be traced to the emphasis on quick achievement fueled by the unquestioned belief in the advantage of the first mover, irrespective of the state of the technology and/or the make. Just like in the case of Net 1000, this belief prompted many business plans to see the facts with rose-colored glasses and throw away caution and risk assessment as the improper behavior of Internet companies.

SUMMARY

Project execution is a continuous process to align the resources and to steer them in the intended direction, taking into account the changes that might be necessary to reflect changes in the environment. Changes in the project specifications while the project is underway are often unavoidable because of many factors: evolution of the technology or the business environment, budgetary instabilities, internal reorganizations, and so on. Successful projects are not only congruent with the company's strategies but have effective scope management procedures. The process entails the following:

- Defining the scope of the project
- Keeping track of the project progress
- Using the scope statement throughout the project for making decisions about change requests
- Tracking changes by documenting the change request in terms of the source of the change, its description, the reasons of the changes, their impact, and the approval chain of authorities.
- Involving the project stakeholders at a given level of authority in the decision regarding the scope changes.

Change control gives the project manager the possibility of tracing the reasons of the changes in the baseline plan as well as their consequences. Without such discipline, the project environment can quickly degenerate with generalized confusion over the project direction. While all projects need some kind of change control, the strictness of the policy depends on the project context.

5

TIME AND COST
MANAGEMENT

The service development organization is usually not the locus of business decisions regarding major investments, such as network upgrades or new services. Instead, it proposes a budget and a schedule and prepares a cash flow analysis that will drive the value analysis. Similarly, while accounting specialists track the actual expenditures on a project, project managers monitor the performance to assess the execution and detect any evidence of latent problems that could conflate into a major threat. This chapter contains a treatment of the basic concepts that guide that intersection between the project and the larger business aspects of telecommunication services.

SCHEDULING

A schedule is a network of interconnected tasks arranged over time to attain the project goals. A useful schedule must:

1. Be objective and self-consistent
2. Be realistic and captures the status of the project
3. Provide an up-to-date picture of the situation on the ground

As a corollary, a schedule serves also as a communication tool among the project stakeholders to gauge progress, detect potential trouble spots, and reach agreements on corrective actions, when needed.

Schedule development will not be treated here because it is the subject of many texts such as Kerzner [1998, pp. 641–685] and Vallet [2003]. Milosevic [2003, pp. 171–221] provides a comprehensive list of scheduling tools; Haughan [2002, pp. 21–23] shows how to adapt the WBS to service projects, while Desmond [2004, pp. 85–111] focuses on projects in telecommunication services. In summary, the various steps consist of: definition of the activities and their sequence, estimation of each activity's duration, resource planning (people, equipment, materials), and budget preparation. Once a schedule has been fixed, a schedule risk analysis can be carried to determine the probabilities of overrunning the end-date and to see if changes can be made to mitigate those risks [Goodpasture, 2002, pp. 81–96]. After a baseline plan has been agreed to, schedule and cost control involves monitoring the progress and tracking all deliverables and milestones. Retroactive changes to the baselined schedules (budgets and costs) are not permitted, except to correct errors. Variance analysis helps in understanding the actual status of the implementation, suggests corrective actions, and helps construct a historical record for the future. Should changes need to be made, they would have to go to through the change control procedure.

The *critical path* is the longest sequential series of tasks from the start to the end of the project. A delay in any task on this critical path ends up delaying the entire project. Therefore, any attempt to reduce the time to deliver the project should focus the effort on compressing the time needed for the activities on that path. Critical path analysis requires an underlying probabilistic structure—that is, to have probability estimates for the various activity durations and to take into account all uncertainties that may arise because some noncritical paths may be become critical under a certain combination of events.

Schedule tracking may be a source of friction with the project team unless its purpose is explained taking into account cultural attitudes toward time and contractual commitments. This should be one of the concerns of project managers.

Delays in Telecommunication Projects

Many delays in a telecommunication service project are due to causes outside the direct control of the project team. For example, they could result from:

1. Lengthy negotiations to get the right of way to access a building or campus to install equipment or lay down cables.
2. Unexpected difficulties in the installation of network equipment, such as in difficult terrains or in remote areas.
3. Weather conditions (hurricanes, floods, earthquakes, hurricane/tornadoes, blizzards, heat waves, etc.) that hinder outdoor activities, such as laying out cables.
4. Work rules on the number of work hours per day, holidays, and so on. Consider the case of laying fiber-optic cables next to a train track. Train schedules need to be considered so that workers can lay out the fiber in between trains. However, it takes about one hour to set up the equipment, and it takes the same amount of time for teardown and clearing the passage before the passage of the train.
5. Logistics (shipping, customs procedures, installation), which are important in the case of international deployment.
6. Unexpected difficulties in getting a stakeholders agreement. For example, upgrades of data networks are typically done during a scheduled maintenance (in contrast with public voice networks, where the service remains available throughout the upgrade). If the network is carrying customer traffic, all customers need to be in-

formed several weeks ahead of time so that they can make their own preparations. If one major customer is not ready, this may force rescheduling the upgrade.

As a consequence, the possibility of delays has to be taken into consideration by preparing contingency plans.

We now present some of the ways for compressing the tasks that are under the control of the project.

Compressing the Schedule

There are three ways of reducing the time for any project:

1. Rearranging the network of activities to allow for parallel activities. Typically, this requires breaking a larger task into smaller components that may start without waiting for all the logical predecessors of the larger task. This technique is called fast tracking in construction projects, concurrent engineering in manufacturing, and rapid application development in information engineering [Rad, 2002, pp. 90–91; Ward 1998, pp. 215–219].
2. Compression of the critical path by adding resources (human and material) without altering the sequence of activities. This action is called *crashing* a project [Ward, 2000, p. 57]. The word makes the reader think of a terrible accident, which in a sense it is, because this is a last resort to attempt to meet the project schedule by adding more resources or increase resource utilization, at the cost of augmented expenditures and reduced efficiency because of the coordination overhead.
3. Reducing the quality of the output.

The set of activities used to compress the schedule is called "crashing," even thought that term gives an impression of an accident. Concurrency requires skilled project management and increases the project susceptibility to perturbations if one subtask has a significant impact on the whole network of tasks. Also, although the purpose of crashing is to minimize additional costs, its effects are confined to tasks that are effort-driven and not duration-driven. For example, if a network verification test requires that error performance by observed for 24 hours, then adding resources will not shorten its duration. Furthermore, there is always an adjustment period when human resources are added to an existing task, the duration of which depends on the skills needed, the social interactions among the individuals, and so on. Finally, the overhead due to additional coordination should be also considered.

Some techniques for shortening the duration of telecommunication service projects are:

1. Re-use of previously tested network equipment and software releases from already certified vendors to shorten the time for installation, even though this may reduce the service features set.
2. Use of manual operating procedures for provisioning and billing until the OSS are able to handle the new service automatically. This is an error-prone solution that does not scale and is useful only for individual cases.
3. Modification of the scope of the project to postpone activities causing delays—for example, by offering only those features that are known to work.

4. In case of site installations, start with sites that are expected to have the highest probability of success. This has the advantage of uncovering problems that might have been missed and resolve them with less pressure and of showing immediate progress to gain momentum. The experience gained and the slack can be used to address problematic installations.

In any event, the risks and uncertainties associated with time saving compromises must be handled properly through contingency plans and a cost–benefit analysis.

COST MANAGEMENT

Cost management is the process of estimating the cost starting with the work breakdown structure and then controlling the expenditures in an environment of continuous change. The idea is to minimize the cost variance—the original estimate may have been inaccurate—while sustaining the project course in terms of schedule and deliverables.

In the phase-wise stage gate approach, a project passes through the several phases with major decision points. During the inception phase, knowledge of the project details is the lowest and the risks are higher, so the estimation is repeated several times based on project-specific data. According to the project management literature, the initial estimate—rough order of magnitude analysis (ROMA) or conceptual order of magnitude—is refined as the implementation advances with a shrinking margin of error as shown in Table 5.1.

Table 5.2 identifies the main cost components for the introduction of a new telecommunications for different types of innovations. The most accurate and reliable cost estimates for a project are those made bottom-up, starting from the lowest level of the WBS, taking into consideration the estimated resource usages and the duration of task. When the historical records are available, modeling tools and techniques can extrapolate past experience to estimate the cost and the schedule, taking into account the properties of the new design.

Thus, for nonincremental innovations, historical data provide limited guidance. Estimates will be based on subjective factors such as technical knowledge, market experience, and the internal dynamics of the organization. In other words, the limits indicated in Table 5.1 may be irrelevant for the following reasons:

1. The technical requirements may be incomplete because of lack of knowledge of the technology
2. Benchmarking data from the industry are not available or they do not exist.
3. The business requirements are not fully specified because it is not clear how the end-user will use the service.

Table 5.1 Guidelines for ROMA Cost and Time Estimates

Subphase	Lower Error Margin	Higher Error Margin
Concept	−25%	75%
Feasibility	−10%	25%
Definition	−5%	10% (0% for time)

Table 5.2 Main Cost Components in Telecommunications Service Development per Innovation Type

Type of Innovation	Major Cost Component
Radical	Development, engineering, testing, marketing training
Platform	Development and engineering and testing
Incremental	OSS development
Architecture	Marketing

4. The belief that the first mover has the advantage takes away any incentive to spend time to get better estimates.

5. Nonrecurring costs to transfer technology, as well as changes needed to correct design deficiencies and to conduct acceptance testing, may be difficult to evaluate.

6. Changes in the market conditions or in the regulation modify the assumptions used in making the business case.

7. Finally, there is a tendency to build an optimistic business case for the introduction of new services. One reason is to ensure proper funding, since most investors prefer rosy pictures. Another is that it is not possible to anticipate all the hurdles that a new technology would have to overcome.

As a result of all these factors, in the early phases of a project that is not an incremental innovation, the planners have a limited view of the total effort needed for system testing, integration with existing systems, and operations and business processes because many unpleasant surprises could occur. In these situations, the real status of the project, irrespective of the initial estimates, would be available at the working level. In other words, knowledge flows bottom-up, from the developers to project management and the sponsors. A necessary condition for this to occur is an atmosphere of trust and mutual respect established by driving out fear as in Deming's point number 7 [Deming, 1986]. This fear factor seems to be confronting NASA management today [Prince, 2003].

PROJECT TRACKING WITH EARNED VALUE ANALYSIS

Project tracking is the process of observing and plotting the parameters describing the monetary and time expenditures in the project execution. Such plots are used to analyze the impact of the variances on the forecast cost and the time to completion and to make adjustments accordingly. This analysis is tied to scope management because when the variance exceeds a predefined threshold, a change request is made to reconsider the budget, the schedule, or the feature set to be delivered. The process of comparison and evaluation is called the *earned value analysis* [Fleming and Koppelman, 2000; Koppelman and Fleming, 2001].

Earned value is an industry standard methodology to track the project performance by integrating the work scope with the schedule and expenditure. It originated from the management complex projects (e.g., defense contracts). Its precursor is a method called Cost/Schedule Planning Control System (C/SPS) used by the U.S. Department of Defense [C/SSR, 1996]. It is now defined in ANSI/EIA-748-1998 and can be calculated by many computer packages. A comprehensive bibliography of the literature is available at http://www.suu.edu/faculty/christensend/ev-bib.html and is maintained by Professor David Christensen, Chair, Department of Accounting and Information Systems, School of Business at Southern Utah University. Other useful sites are available at www.acq.osd. mil/pm and at http://www.goldpractices.com/practices/tev/index.html.

While the calculation of the earned value can be an effective tool in measuring progress, it requires planning and budget allocation upfront. In addition, guidelines are needed to measure the project progress so that whenever a milestone is reached, a value is earned in proportion to the total allocated budget of the project. The total value earned so far is then compared to what was expected, thereby giving an overall indication on the direction of the project.

Metrics for the Earned Value

Earned value measures the way the resources are mobilized to meet the project's goal. Even though the budget is usually represented by a monetary value, from a project management perspective, resources in terms of equipment and expertise are as equally important; thus hours of work could also be a suitable measure. ANSI/EIA-748-1998 specifies the following three methods for evaluating the value of the task being executed: the *discrete effort* method, the *apportioned effort* method, and the *level of effort* method.

Discrete Effort Method. In the discrete effort method, the earned value is based on the completion of discrete tasks. The value is computed using valued milestones, standard hours, or management assessment. Exit criteria should be defined for each task to make it easier to track the task completion and reduce the subjectivity as much as possible.

In the *valued milestones* approach, the earned value is a binary variable (done or not done) that can be integrated with the major decision process through the quality gates of the service realization process. The standard specifies two rules for assigning value:

- In the 0/100 technique, the project is credited 100% of the budget value when the work is delivered (the task is finished and the successor task begins).
- In the 50/50 technique, the project is credited half of the budget for the task after the start and the rest upon delivery.
- The following variations are sometimes used, even they are not specified by the standard [Rad, 2002, p. 75]:
  - The 20/80 technique: After the start, the project is credited 20% of the value with the rest upon delivery.
  - The 30/30/40 technique: The project is credited 30% after starting, 60% when as it progresses, and 100% at completion.

In the *standard hours* approach, the value is measured as the cost of staff in terms of work hours. For new developments, the cost of expertise is often the predominant component of the cost [Thoren, 2000].

In the *management assessment* approach, the completion of a task or a group of tasks is given an earned valued based on its relationship to the total work to be done. For example, the earned value in the acceptance testing part can be made proportional of the number of tests executed so far.

Apportioned Effort Method. The apportioned effort method is useful for effort that cannot be directly measured or divided into discrete parts but is related to other measured work so that the earned value increases each time one of the precursors are accomplished. For example, starting of acceptance testing of a network element depends on the progress of the supplier, so that when the supplier meets one of its milestones, the earned value for the corresponding task increases.

Level of Effort Method. The level of effort method is useful when activities cannot be subdivided and measured such as in the case of fundamental research problems. In such a situation, it is not easy to subdivide the tasks and anticipate milestones. The earned value in this case is based on the passage of time.

Budget Types

Once the budget has been identified, it is divided into several categories

1. Contract Budget Base (CBB): This is the total contract cost minus a contingency to cover emergencies.
2. Management Reserve (MR): This represents an amount withheld for management control purposes—for example, to account for tasks that are defined in the statement of work but that have not been planned yet. This amount is typically less than 10% of the overall budget.
3. Undistributed Budget (UB): This is a place holder for the amount that could not be allocated at the start of the project.
4. Budget at completion (BAC): This is a budgetary number representing all authorized work as defined in the statement of work. This number cannot change unless the statement of work is changed. It is also called the performance measurement baseline.

Monitoring Project Progress

The method selected to credit the value should be clearly articulated in the project plan and used consistently. The amount of progress is then computed by summing the progress in each of the subtasks that constitute a major task in terms of the total earned value credited for the project for executing that task.

The difference between budgeted value of the work accomplished and the expenditure evaluates the cost overruns or *cost variance*. A second comparison concerns the gap between the planned spending for the amount of work already *done* and the planned spending for the work that *should* have been done were the baseline schedule adhered to. This second comparison gives the potential delay or the *schedule variance*.

The following parameters are used to generate an integrated set of measures of the project progress:

1. The budgeted cost of the work scheduled (BCWS). This is a time phase budget spread of the required resources to complete the task. It forms the baseline for measuring the actual performance (performance measurement baseline).

2. The budgeted cost of the work performed (BCWP). This is the earned value that represents the current status of the work in terms of the cumulative expenditures for performing the work so far using the values allocated in the baseline schedule.

3. Actual cost of work performed (ACWP) is the value of the cumulative costs incurred to accomplish the work performed within a given time period (also known as actuals).

Figure 5.1 shows the budget values and primary measures used in the earned value analysis.

Measures of Efficiency

The efficiency of the work is assessed using derived measures from the primary measures. These derived measures are as follows:

- Cost variance CV = BCWP − ACWP. This is the numerical difference between the earned value and the actual cost; that is, it represents the deviation of the actual from the budget in accomplishing the amount of work already done.
- Schedule variance SV = BCWP − BCWS measures the difference between the amount of work accomplished in a given period with the amount of work planned for the same period.
- Estimate to complete (ETC) is the budgeted cost to complete the remaining work and is computed as BAC- BCWP.

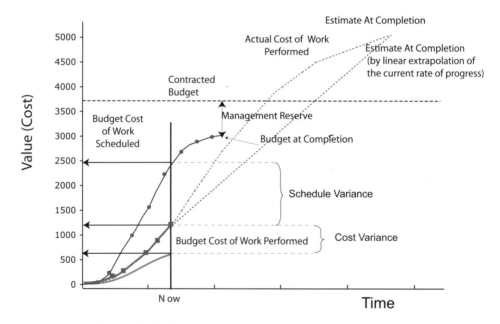

Figure 5.1 Budget values and primary measures of earned value.

- Estimate at completion (EAC) is the final expected cost to perform all the work agreed to:

$$EAC = ACWP + ETC.$$

EAC is one of the key indicators of the status of the project. It is worth noting that studies of hundreds of programs have revealed that a good indicator of the lower bound of the cost at completion can be obtained after completion of about 15% of the contract work [Sparrow, 2000]. Of course, whenever the scope or schedule change, the assumptions that are the basis for the EAC should be reexamined and the estimate updated accordingly.

Two indices can be derived from the primary measures to assess the status of the project using normalized parameters. These are as follows:

- The schedule performance index = BCWP/BCWS. This index is < 1 if the work performed is less than what was scheduled, and it is ≥ 1 otherwise.
- The cost performance index = BCWP/ACWP. This index is < 1 if the work performed is consuming more hours than originally planned, and it is ≥ 1 otherwise.

In general, the CPI does not improve significantly once 15% of the contracted work has been accomplished. Therefore, at this 15%, the value of CPI can indicate that some changes in the budget or schedule are warranted. For example, if the value derived from the data at the 15% mark is 0.65, this suggests that only 65% of the work will be completed at the scheduled project time so adjustments are needed in the budget, schedule, and/or the scope of the work.

Prerequisites for Earned Value Analysis

The proper operation of the earned value method depends on four pillars: a fully defined scope, a bottom-up WBS, a detailed schedule, and a protocol for data collection, tabulation, and dissemination. In the case of a rolling wave method of development, such as for nonincremental innovations, the project starts without the WBS and the schedule being fully developed. Quite often, the detailed schedule can be defined only 2–3 months in advance because there are many contingencies. Thus, tracking of earned value during project execution is applicable at the project level in the case of incremental innovations; for other types, it can be done at a local or functional level as shown in the next example.

Earned Value Analysis in Telecommunication Projects

This example treats the case of certification of a new network element that implements a new technology to be verified before deployment in the network. Because of the lack of experience with the technology and its embodiment, it is not known which areas could be problematic and how much time effort would be needed for each area. The measurement of value will follow the discrete effort method; accordingly, the graphs in Figure 5.2 depict the total hours scheduled for testing, the actual hours spent in the project, and the actual hours conducting the testing. These will be used to measure the schedule cost, the actual expenditure, and the actual work accomplished. We see in this case a confirmation of the rule that with 15% of the work executed, a linear prediction can give a reasonable estimate of the time to completion as well as the cost to completion. From the project man-

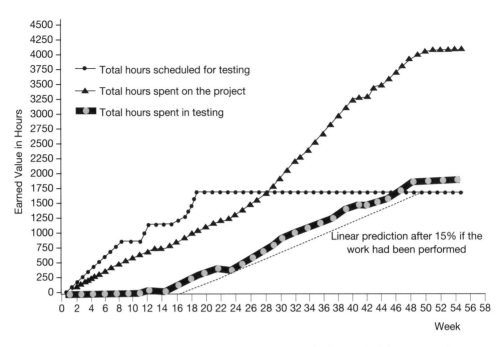

Figure 5.2   Tracking of testing using the earned value methodology.

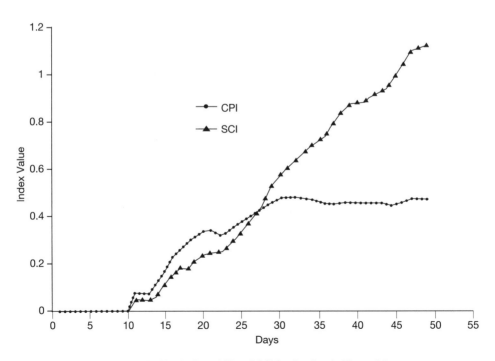

Figure 5.3   The indices CPI and SCI for the data in Figure 5.2.

agement viewpoint, we see that the testing effort was correctly estimated while the calendar time underestimated. There are two reasons for this error: The ramp-up time was longer than expected and the original estimates did not include the related overhead (e.g., training, test case preparation, equipment setup, meetings with the vendor, etc.).

Figure 5.3 shows that the CPI and SPI indices. We see that, for this project, the CPI has been stable for the second half for the project. The CPI is around 0.5 because the work performed consumed twice as many hours as originally planned (due to other activities that the testers were involved with).

SUMMARY

Schedule and cost pressures originate from demands by clients or executives based on competitive needs. The purpose of cost and time management is to follow and report the real project status, at regular intervals to provide early and accurate warnings when it is still possible to resolve problems without incurring large expenses. With such a data and in an environment where discussions are encouraged, it would be possible to converge on the best course of action given the actual conditions. On the opposite side, demands that appear to the members of the project group to be arbitrary have negative impact on the morale of the project members, the overall cost of the project, and the overall quality of the end delivery.

6

INFORMATION AND COMMUNICATION MANAGEMENT

Communication management is the set of policies and procedures to ensure that project data are collected, processed, and made available in the form of useful and timely information to those who may need it in the course of their work. Thus, it covers the prescription of processes for generating, collecting, distributing, updating, and storing project-related data and documentation in a standardized manner and ultimately disposing of project related information. Part of these processes relate to a project in terms of authorization and access privileges. Another aspect is its contribution to current and future projects in the form of knowledge management.

The focus of this chapter is on the internal communication among the stakeholders. Excluded are the communication interfaces with regulators, financial institutions, standard bodies, or the general public. We consider the effect of cultural preferences on the types of communication channels for a project. Finally, we suggest a few measures for the effectiveness of project communication.

THE ROLE OF COMMUNICATION MANAGEMENT

The main goals of communication management are to

- Facilitate the dissemination of information
- Ensure team cohesion
- Build a historical database

The relative importance of each of these items depends on the innovation type. Direction and control dominate in the case sustaining innovations because the project environment is stable. In disruptive innovations, the emphasis is on knowledge sharing, team cohesion, risk mitigation, and uncertainty reduction. Another goal is to build a repository of historical records so that future project teams could learn from the experiences gained.

We now consider each of the three items.

Dissemination of Information

Project information must be disseminated in an efficient and timely manner to improve the coordination among parallel activities and to allow early identification of unknown interdependencies, thereby helping risk mitigation strategies. Project teams usually generate large amounts of raw data or documents. In a flat organization, the number of communication channels among N entities is $N(N - 1)/2$. In disruptive innovations, new knowledge is generated at a rate that can overwhelm hierarchical distribution channels. At a minimum, the establishment of a common vocabulary reduces confusion over the work content, its deliverables, and the expected quality.

In the absence of a systematic policy for communication, the project information is exposed to misinterpretations due to accretions, omissions, or commissions, with dramatic consequences. The failure of the Mars Climate Orbiter (MCO) spacecraft to properly orbit Mars was due to miscommunication between the Jet Propulsion Laboratory (JPL) and the contractor Lockheed Martin. The JPL people thought the navigation data were expressed in metric units, while the Lockheed Martin personnel assumed them to be in English units. When time came for the final orbit insertion burn, the spacecraft was not in the place where it was supposed to be, so either it came to close to the planet and burned in the Martian atmosphere or it was too far and continued past the planet [Stephenson, 2002].

To summarize, from the perspective of knowledge creation and acquisition, the objectives of communication management are as follows:

1. Improve the exchange and sharing of pertinent data among the project team members.
2. Spread the knowledge to all who need it so that it is not confined to a small group.
3. Convert local knowledge to more general principles that can be used in the current and future projects.
4. Abstract the knowledge gained during the project implementation to be a source of competitive advantage.

Team Cohesion

Part of the information flow within the project team concerns the coordination, motivation, and explanation. Confidence and trust are social constructs, therefore team cohesion is improved when the agendas of the various parties are understood and taken into consideration by all stakeholders. In addition, the sharing of explicit and implicit assumptions helps in building a stronger collaboration based on common expectations (i.e., things that are taken for granted). This commonality, in turn, defines the framework within which acceptable solutions are negotiated. Common planning sessions improve team dynamics through a better understanding of the methods and procedures of each organization.

To establish effective communication within a project team, a powerful and compelling narrative must be shared. Without such a common vision, it is not realistic to expect the formation of "self-directed teams." The success of virtual teams depends in part on efficient exchanges among team members across multiple geographic areas, even though they may report to different authorities or have a predisposition toward one form of communication over the others.

Historical Database

Project repositories provide a coherent way of storing project documents that can be used for future projects. Such a corporate memory, when properly used, can be a source of competitive advantage for similar projects or a stimulus for innovation. The historical database is more important when team members are distributed or when team membership varies significantly.

Building a historical record is difficult when quick and dirty techniques are encouraged, when the goal is to learn quickly but not necessarily well, or when there are constant reorganizations. Another difficulty is the belief that hoarding information confers power. The communication plan helps in overcoming these hurdles by emphasizing the project rules regarding communication and allocating a part of the project budget and effort toward that goal.

COMMUNICATION AND OUTSOURCING

There are several structural incentives for outsourcing, particularly in the case of telecommunication services [Awde, 2004; Rad 2002, p. 99; Roche, 1998]:

- To ensure round of the clock operation of customer care centers and maintenance by spreading the operation in locations at different time zones. This requires the segmentation of the work in chunks that can be managed independently, their documentation, the documentation of the requirements, and the definition of the quality measure for the deliverables.
- To reduce the time to market by distributing the work according to the available competence rather than bringing that experience in house.
- To supply missing expertise in non-core functions, particularly because many telecommunication operators are no longer vertically integrated.
- To migrate labor-intensive functions in the service to areas with educated and skilled staff at lower costs.
- In the case of global telecommunication services, to have the customer care functions closer to the end-user,

As an illustration, consider a growing trend among some telecommunication operators to outsource the management of their networks to their suppliers or to third parties. One example is the accord between Hutchinson Telecom (an operator from Hong Kong) and Ericsson to run the first's Australian wireless network. The managed infrastructure comprises second-generation (2G) equipment supplied by Samsung and third-generation (3G) nodes from Ericsson itself [Brown-Humes, 2002]. Such an outsourcing accord introduces additional managerial layers to routine operations such as planning of a network upgrade or responding to a trouble report. Clearly, outsourcing network management calls for a reliable telecommunication infrastructure, strong communication processes and the alignment of business processes among the two parties of the contract.

Conversely, the inadequacy of communication management can engender unfavorable consequences in outsourcing situations such as:

- The disclosure of proprietary information during training of the contractor's staff to perform the outsourced tasks
- Loss of expertise unless there is a systematic method for acquisition of knowledge from network events
- Lock-in of the contractor because of the sunk cost in training and in establishing the communication infrastructure
- The need for frequent face-to-face meetings for coordination and conflict resolution

In other words, when the outsourced function is critical to the success of the service, knowledge acquisition and diffusion should be an integral part of the plan to monitor and track the contractor's performance. This is why outsourcing is a strategic step that should not be undertaken lightly in the hope of a quick fix and without consideration of all the tangible and intangible costs associated with the outsourcing decision.

THE COMMUNICATION PLAN

The communication plan describes the formal mechanisms, routines, and procedures for information coding, exchange, protection, and storage, taking into account the project organization. Communication effectiveness is a function of matching the processes and the protocols with the nature of the project team and the organization design. The more complex and formalized the organization, the more formal the communication plan.

Among the items the plan will have to address are:

- The type of data to be acquired, when it should be collected and by whom
- Classification of documents in terms of confidentiality
- Access and security policy
- Architecture for storage, distribution, and archival
- Technical aspects of the access—that is, the type of terminals, the access link, and so on.

One possible way of approaching all aspects of communication management at once is to consider three factors: the audience, the circumstances, and the nature of information, as shown in Figure 6.1.

Audience

For efficiency and information protection, the information presented to the stakeholders should give them what they need and at the appropriate level of detail. Typically, corporate managers are more interested in the project risks (schedule and cost) or the opportu-

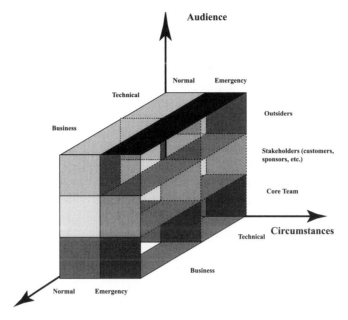

Figure 6.1   Considerations in communication management.

nities for additional gain through patents, trademarks, copyrights, and licenses. The project team, in contrast, focuses on the tactical aspects such as technology and system performance. Vendors, suppliers, and contractors have, for all intents and purposes, similar information needs as the technical organizations. Support organizations (procurement, finance, human resources, legal, etc.) seek the data sets that assist them in the execution of their duties.

Circumstances

The communication plan has to take into account normal operation as well as exception conditions such as escalations or emergencies. The normal procedure is to update key information at regular intervals to allow tracking of the progress. When the delay in a scheduled activity exceeds a predefined limit or puts the whole project in danger of not meeting its end date or missing a major milestone, an escalation procedure can be invoked. Finally, the communication plan must also define how to resolve conflicts or face emergencies.

Nature of Information

In telecommunication services, the project documentation typically includes marketing development plans for the service, technical requirements, engineering guidelines for the network, test plans for equipment certification and turn-up in the network, responsibility matrices, business flow processes, budgets, cost estimates, reference manuals, white papers from manufacturers, project minutes, status reports, tracking issues lists, defect lists, training information, laboratory tools, approved changes, customer letters, legal documents, customer feedback, lessons learned, trip reports, and so on. Standard templates help ensure that the information is communicated in a uniform manner.

Content of the Plan

The contents of the communication plan specify items such as

- The needs of each audience type (i.e., what)
- The responsibility for information generation and distribution (i.e., who)
- The periodicity of the information generation distribution (i.e., when)
- The templates for the various documents
- The mechanics of the distribution (i.e., how)
- Criteria for information prioritization (i.e, why)
- Escalation procedures for problems that may affect the project
- Procedures for interfacing with vendors, internal or external customers, and so on.

One possible tool for arranging these data visually is to use one or several communication matrices. Table 6.1 displays one such communication matrix to identify the list of invitees to various project meetings.

COMMUNICATION CHANNELS

The channels and processes used for communication will depend on the objectives pursued, the target audience, and the security policies of the project. As projects become global in terms of scope or team membership, linguistic and cultural differences become more important; a formal plan is needed to encourage open communication and overcome all the barriers to information exchange. However, projects that have a global dimension, in terms of scope or of team membership, linguistic and cultural differences among team members may influence their preference for one way of communication over the other as well as their understanding of what to communicate and how [Chevrier, 1996; Minor, 1999, Sherif, 2000; Sherif and Pérez, 1999]. There is a wealth of factual information on the issues that arise in cross-cultural communications that human resources specialists as well as international business consultants have identified as factors in multinational companies, in international negotiations, and in global projects [Breuer and de Bartha, 1993; Trompenaars and C. Hampdend-Turner, 1998]. With the increase of offshore development, more attention is being now given to the impact of culture on global software projects [Carmel, 1999; Gezo

Table 6.1 Communication Matrix for Meeting Attendance

Name	Project Status Meeting	Weekly Conference Call with Operations	Vendor Meeting	Joint Program Meeting
ABC	✓	✓	—	—
MNOP	—	✓	—	✓
...	...	...	...	...
XYZ	✓	—	✓	—

et al., 1999]. Contentious styles of communication must be kept to a minimum, keeping in mind the cultural contexts so that the team can operate smoothly. For example, because the notion of urgency varies from culture to culture, communications about deadline and the seriousness of missing commitments must be explained clearly by showing their impact on the overall project schedule (i.e., it is not sufficient to say it verbally or to repeat words that may be misunderstood). In other cases, the definition of the standard of expected performance and development and how the tasks will be carried will have to be in excruciating detail. Thus, exhortations to "promote trust among team members" or to abolish "intimidation tactics, blaming, harsh language, or humiliation" [Holahan and Mooney, 2004] cannot be separated from the cultural context, because the boundaries between constructive and destructive criticisms vary among cultures.

Whether communications will be verbal or written depends on the cultural attitude to the written word. In some cultures, written documents are reserved for binding commitments. Furthermore, there may be a certain hesitation to exhibit individual knowledge (or wealth), besides the fear of making mistakes of style or grammar, which translates in the amount of time needed to write documents. Some of the communication channels are:

1. One-on-one communications
2. Meetings
3. Telephony and teleconferencing
4. E-mail
5. Intranets and project portals

In general, collaboration tools must be congruent with the organization's competitive and individualistic culture as well as its reward system [Orlikowski, 1996].

Let us evaluate each channel.

One-on-One Communication

One-on-one meetings are necessary to accomplish the most difficult tasks of the project or when political and technological uncertainties are high. In high-context cultures, as defined by E. T. Hall [Hall and Hall, 1990], a minimum of two face-to-face meetings per year is essential for maintaining a good working relationship, particularly when people are expected to interact on a regular basis [Asselin and Mastron, 2001, p. 191].

Meetings

This a common approach to exchange information, to establish personal relations, or to reach decisions. Meetings can be effective if they are prepared with care. There are a number of established techniques for running effective meetings that will not be reviewed here. We highlight some aspects that would be of interest in telecommunication services projects.

Depending on the purpose of the meeting, the agenda can be rigidly defined or can be more flexible. The frequency of each type of meeting and the way decisions are reached depend on the culture of the organization and/or the society. Also, the way it is conducted would determine whether the meeting should be face-to-face or could be conducted remotely.

Some particular meetings are as follows:

1. *Project Kick-off Meetings.* The purpose of a kick-off meeting is to formally announce the start of the project and to define the broad areas of roles and responsibilities.

2. *Status Meetings.* These are typically periodic (e.g., weekly) events to keep the internal stakeholders abreast of current schedules, issues, progress or changes of scope, technical requirements, or specifications. Other agenda items include resource gaps, potential process enhancements, organizational changes, funding shifts, and system enhancements. In a large deployment, such as in a network upgrade, project meetings involve many participants spread over a large area: engineering teams, on-site work force in work centers, members of the organizations involved in provisioning, facility planning, network maintenance, customer care, product management, account teams, and so on. Such meetings are now held via audio teleconferences.

3. *Working Meetings.* These are conducted to identify and resolve technical problems or to report on technical results or to prepare a subphase of activity. Progress meetings should be take place periodically (e.g., weekly) with minutes recorded and distributed.

4. *Situation Room (War Room) Meetings.* These occur during high-risk situations such as during network upgrades or in network outages where the people involved can brief each other and exchange up-to-date information with each others or across shifts.

The communication plan must specify the various types of meeting, their frequency, their objectives, and the roles of the attendees. Meetings for dialogue are one important way to establish a consensus by exposing and analyzing various viewpoints. Dialogue is a good way to form a common understanding, and this is why it is one effective tool for conflict resolution.

The actual running of the meeting is also a cultural phenomenon. In cultures that Hall [1976] defined as high-context or diffuse, things to do not have to be spelled out explicitly and meetings are a way to exchange information through facial expressions, body language, or seating arrangements. In such an environment, interpersonal contacts are important to avoid misunderstandings. In contrast, in low-context cultures, the information is explicit. In these cultures, meetings take place in a linear fashion with a specific agenda to achieve results that can be *reported*. There is little room for "open space" and for discussion outside the subject of the meeting. Statistics on newspaper readership and sales confirm that the population of Northern Europe (Finland, Germany, the Netherlands, Sweden, United Kingdom), which is low-context and monochronic (one thing at a time), relies more on the press for information, while more to the south, such as in Belgium, France, Greece, Italy, Portugal, and Spain, where the culture is high-context and polychronic (capable of multitasking), people get more of their information from radio and television [*Futuribles,* 1999].

Similarly, the order of presentation is cultural-dependent. Three possible ways that are in used are:

- General → specifics → conclusions
- Specifics → general → conclusions
- Executive summary → general → specifics → conclusions

Of course, being aware of these cultural differences is essential if the project team assembles people from different cultures. More generally, team members should be aware of the hazards of communication and that decoded messages may have unintended significances. In other words, people should verify that they have been correctly understood and interpreted before coming to conclusions on how to respond.

Telephony and Teleconferences

Although remote conferences (audio and video) are common nowadays, their effectiveness depends on the amount that the various parties spend in preparation. Cultural differences affect the quality and the spontaneity of the exchanges in many ways; furthermore, the characteristics of the telephone conversation is affected by culture—for example, the percentage of time each party is talking (speech activity factor), which lies between 30% and 40%. Whether a telephone call or an e-mail is the preferred method of collaboration varies according to culture as well [d'Iribarne, 1998]. Obviously, the purpose of a meeting and the way it is conducted will determine how remote conferencing should be managed.

E-Mail

As can be anticipated, e-mail and written correspondences are preferred in low-context societies, while phone conversations and direct meeting are typical of high-context cultures. Furthermore, although English is now the de facto international working language for communications, misinterpretations are more likely to occur because of the "lack of attention from the nonnative speakers to the metatextual aspects of the text as a discourse event, such as the authors' purpose and his target audience [. . and] also from the nature of the text itself . . . [The] mental representation of the texts [by nonnative speakers] is fragmented, incomplete and inappropriate" [Daoud, 1991].

Intranets and Project Portals

Web tools are available to manage the project documentation as well as project performance: project agendas, meeting minutes, technical plans, time line, vendor documentation, test results, and so on. On-line systems can reduce the cost of publishing common project documentation, provided that they are well designed—that is, a user-friendly interface, accurate and up-to-date catalogues, cross-referencing of the documents as well as easy searches for retrieval.

A virtual project library does not always replace a physical library completely. The latter should contain a subset of the documents to preserve the data for the future (once the electronic storage system is retired). In addition, using Web access requires a strong security architecture and security policies to protect the information.

Security policies cover several areas such as

- The physical security of the site as well as physical access to the server
- Access control to counter the threat of unauthorized operations
- Protection of the confidentiality and integrity of the project data
- Protection against attacks
- Intrusion detection

- Audits and reports
- Containments
- Recovery including backup policies

EVALUATION OF THE COMMUNICATION PROCESSES

Measure of Communication Effectiveness

The measures for communication effectiveness can be objective or subjective. Objective measures include:

- The time interval between the instant when an issue is raised and when it is closed or resolved
- The idle time because of lack of predecessor completion
- The frequency or reworks because the information did not arrive on time
- The percentage of missing documentation
- The blocking probability and the existence of single points of failure

Subjective measures can be obtained from internal surveys of the team members' attitudes or by the prevalence of rumors and leaks. Rumors indicate that people do not have the information they need so they resort to guessing. Leaks fall into two categories: the deliberate and the careless [Martin, 2005, p. 128]. Deliberate leaks increase whenever the team is not cohesive or if the legitimate concerns of some stakeholders have not been addressed. Careless leaks occur when the confidentiality policy has not been well communicated to the project team.

Signs of Communication Problems

Without conducting a formal measurement program, casual observation can also point to communication problems. Some of these telltale signs are as follows:

1. The proliferation of acronyms or imprecise vocabulary (same word used for different concepts, or many words for the same concept) in the project documentation can lead to misunderstandings.
2. Lack of common understanding of the magnitude of changes in the scope or in the schedule.
3. Overlapping areas of responsibilities leading to frictions and/or power struggles.
4. Incomplete, missing, or inconsistent documentation:
   - Design documents are not communicated in time to the various responsible organizations.
   - Ambiguous or incorrect documents.
   - Key decisions are not adequately documented.
5. Process inefficiencies:
   - Meetings canceled or rescheduled due to lack of attendance by necessary attendees (functional areas not adequately represented).

- Meetings expand beyond their scope to complete work.
- Frequent interruptions of ongoing activities due to last-minute changes.

Barriers to Successful Communications

Some of the barriers to successful communication are as follows:

- The communication plan is not adapted to the organizational design: In functional organizations, for example, the functional managers have both hierarchical and information powers and influence the way project members receive or share information. Implementing the communication plan has some costs, monetary and operational. Thus, in a matrix or projectized organization, more effective channels are needed to ensure that information can reach its intended recipients rapidly and accurately.
- In cross-functional teams, the specialized vocabulary of the various disciplines is used without explanation. In particular, the excessive use of acronyms can prevent understanding. Because groups tend to evolve their own language depending on their cohesiveness and experience, the plan may be revised periodically during the project progress.
- Inadequate security policy: too lax or too restrictive.
- Inadequate protection against information loss due to the withdrawal of particular individuals.
- Lack of agreement on what information is useful and what is not.
- Information chatter that congest the information channels.

SUMMARY

In a project environment, a large volume of data and documents is being generated, so a communication plan is needed to clarify how this information will be managed. Empirical studies show that there is no single communications media to provide the optimum channel for all types of communications at all times. Different projects need different communication plans depending on the type of the innovation and the project organization. Cultural factors affect the communication management plan in at least three aspects: the choice of the collaborative technology, the social network that produces or consumes the information and the content of the information.

Given the sheer volume and complexity of information, it is important that the communication plan be well constructed by taking into consideration the needs of the various stakeholders. This avoids dumping of information in such a way that the important issues will be lost.

7

RESOURCES MANAGEMENT

Telecommunication services, like all services, are human resources intensive; therefore, the recruitment of the project staff can have an effect on existing services. This is why the management of human resources for projects in telecommunication services cannot lose sight of current needs because, in all but a completely project-oriented organization, the team members continue to report to their functional organizations. In particular, when a project relies on experience garnered from the field, such as for incremental or architectural innovations, the formation of the project team can stretch the available resources to the point of perturbing day-to-day operations.

The presentation in this chapter will summarize a few key points that help the management of n telecommunication services projects. Further details are available in the abundant literature on leadership, team-building, motivation, and conflict resolution. A good starting point is available in the Human dimension track of a comprehensive guide edited by J. Knuston [2001].

FORMATION OF THE PROJECT TEAM

Matching a person's technical skills and personal characteristics to the needs of the project is a major part of team selection and role. Various objective criteria can be arranged in the form of a skill inventory matrix to assist in the selection of the team members. This is an inventory of the skill sets that are needed for each major task of the project, including (a) the technical and administrative aspects and (b) the characteristics of the resources under consideration in terms of technical ability, availability, and so on. A responsibility matrix documents the names of the individuals retained, with one person accountable for each task. When other names are assigned to the same task, they will have a supporting role, perhaps to offer skills needed for task completion or to participate in the transfer of knowledge to build some redundancy. Such a redundancy is a way of mitigating risks due to the unavailability of the resources (i.e., to avoid single points of failures in the project plan). Finally, once the team has been decided, a stakeholder matrix identifies the overall team—the core group and the external members such as customers, vendors, contractors, and so on—as well as any additional stakeholders that can influence the outcome of the project such as regulators, standard organizations, and so on.

Technical ability is not enough to guarantee that the project team will be able to work together. Depending on the kind of people that make up the project team, the work may be enjoyable or painful. The formation of cross-functional teams is the basis of rapid application development (or concurrent engineering), a technique used to shorten the delivery intervals by enhancing communications across divisional boundaries to improve the transfer of complex technologies. Nevertheless, some effort is needed to make sure that people assembled mostly for their technical expertise can work together as a team.

We turn now to the profiles of people that would be suitable to work in a project environment. The discussion is based on Wideman's study [Wideman, 1998] using the Myers–Briggs-Type Indicator (MBTI). The MBTI has been in use for over 50 years as a tool for management training in most large U.S. corporations. Despite the limitations of psychological indicators, particularly if they are applied automatically to an individual, they reveal general tendencies and offer useful insights for management purposes.

The basic postulate of the MBTI is that four separate (albeit interrelated) personal preferences or natural tendencies can define a personality as shown in Table 7.1 [Myers et al., 1998]. These preference categories define individual styles in perception, decision-mak-

Table 7.1   MBTI Preference Categories

Preference		Opposite Preference		
Category	Description	Category	Description	Meaning
Sensing (S)	Relies on practical facts or experience of life	Intuiting (N)	Sees the possibilities, patterns, relationships, and meanings of experience	This a *perceiving* function related to the way people perceive the surrounding and gather data
Thinking (T)	Prefers objective and analytical factors	Feeling (F)	Prefers subjective factors	This a *judging* function regarding the way people evaluate the data that they have gathered
Judging (J)	Prefers organized and orderly ways as well as linear thinking	Perceiving (P)	Prefers to live in a spontaneous, flexible way, aiming to understand life and adapting to it	This is a measure of the degree of *flexibility* when people act upon their decision
Extroverted (E)	Prefers the outer world of actions, objects and other people	Introverted (I)	Prefers the inner world of concepts and ideas	This is a measure of the way people express their conclusion or direct their energies

ing, behavior, and affect: that is, they relate to how people gather information, process them, act on them, and then express their feelings.

These preferences form 16 distinct "personality types" as shown in the MBTI grid of Figure 7.1 [Wideman, 1998], with the percentages showing the approximate proportion of each type in the total U.S. population. Table 7.2—illustrated in Figure 7.2—shows that the U.S. population is heavily skewed toward the sensing preference, particularly the SE quadrant. In other words, about 76% of the U.S. population prefer to gather data with the senses without looking for possibilities, meanings, and relationships among various things [Kroeger and Thuesen, 1988, p. 25]. The least concentration is in the IN quadrant that characterizes recluse thinkers, with the combined preference of introversion and intuitiveness. It is worth noting that numbers are skewed in at least two ways. First, they are based on groups of high school and college students. Second, they did not represent adequately the minority cultures in the United States [Kroeger and Thuesen, 1988, p. 25,

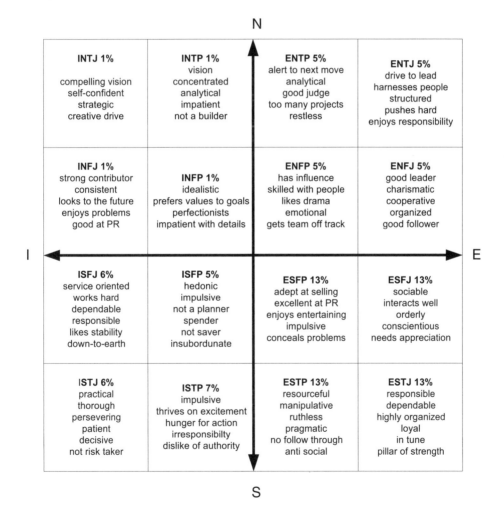

Figure 7.1 The MBTI grid for the U.S. population (the percentages show the approximate proportion in the population).

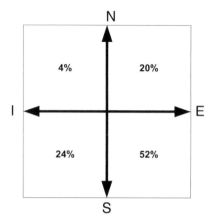

Figure 7.2   Quadrants of the preferences in the U.S. population.

note]. Nevertheless, they clearly show that preferences are not uniformly distributed throughout the population, which is the main point of this discussion.

Note: For the same North American population, Wideman [1998] has slightly different numbers than those shown in Table 7.2.

According to Wideman [1998], the MBTI grid can help identify the types suitable for project work using the criteria of Table 7.3, which are illustrated in Figure 7.3. The conclusion of this analysis is that about half of the U.S. population is not comfortable with project work (26% + 0.5 × 49% = 50.5%).

We can generalize this conclusion to other populations by stating that people recruited solely on the basis of technical capabilities and skills are not always suitable for a project environment—that is, whenever a purely functional organization is not used. In a balanced matrix arrangement the project manager decides the content and the timing of the task execution while the functional manager decides the resources and how things will be done. One of the differences between a functional organization and a matrix organization

Table 7.2   Distribution of the Preferences in the U.S. Population

Preference		Opposite Preference	
Sensing (S)	76%	Intuiting (N)	24%
Thinking (T)	51%	Feeling (F)	49%
Judging (J)	50%	Perceiving (P)	50%
Extroverted (E)	72%	Introverted (I)	28%

Table 7.3   Suitability to Project Management

Suitability for Project Work	MBTI Type
100% suitable	INTJ, ENTJ, INFJ, ISTJ, ESTJ
50% suitable	INTP, ENTP, ENFP, ENFJ, ESFJ, ESTP,ISFJ
EUnsuitable	INFP, ISFP, ESFP, ISTP

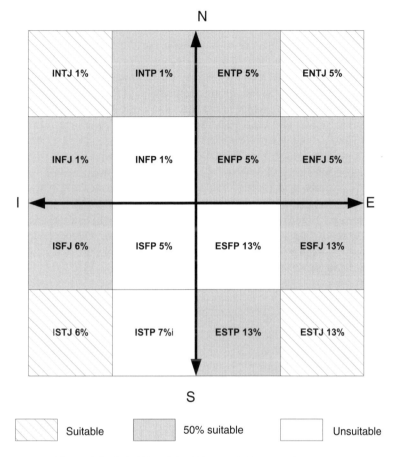

Figure 7.3 Suitability of the U.S. population to project work.

is that in the latter case, team members may find themselves caught in between conflicting requirements and with constraints that are impossible to meet. In a project-oriented arrangement, people have to seek new positions as the project is winding down. In some cases, the functional department where they came from may no longer have suitable positions. Even when team members return to their original functional areas, reentry is not merely going backwards in time to the point where the trajectories have diverged. The reason is that the functional environment has continued to evolve during the project life with turnover of people, changes in work routines, advancements in the professional specialty, and so on. Returning members will have to spend some time to catch up and to adjust to the organizational changes, to update their skills and try to find ways of exploiting their newly gained experience within their functional environment [Sorti, 1997]. This shows the need for assisting the human resources when they transition in and out of projects, at least in their first assignment. This should be under consideration if the project manager does not always get to select the project team, or when the team composition changes as the project goes through its life cycle.

We focus now on the team building activities.

TEAM BUILDING

The output of a successful project team is more than just the sum of contributions for individual members. Team building is the process by which a group of people selected mostly on the basis of their technical expertise can be brought to form a coherent group that shares a common vision, have a strong commitment to that vision, and collaborate to work together to achieve success. A *jelled team* is a team whose members are strongly knit together in a single unit taking advantage of the synergy among them and whose members derive more enjoyment from their work than from the nature of the work itself. Part of the enjoyment comes from the interactions themselves as the members work to overcome the obstacles they face; in other words, the connections themselves are part of the pleasure [De Marco and Lister, 1987]. Such a state was described by the Chinese strategist, Zhuge Liang of the third century A.D.: "The Tao of military operations lies in harmonizing people. When people are in harmony, they will fight naturally, without being exhorted to do so" [Cleary, 1988, p. 8]. In today's management jargon, such teams are called "high performance" or "self-directed" because the group operates almost autonomously [Thamhain, 2001, 2003].

Team Building and the Hierarchy of Human Needs

Maslow [1970] hypothesized a hierarchy of general needs that drive human motivation and that must be satisfied before a person can act unselfishly. Accordingly, the needs at the lower levels must be fulfilled before a person can focus on the higher levels.

1. The first needs to be satisfied are physiological such as air, water, food, sleep, sex, and so on. These feelings push us to alleviate them as soon as possible to establish homeostasis.
2. Safety needs come next. They have to do with establishing security, stability, and consistency in a chaotic world.
3. Love, belongingness, and social integration are next on the ladder,
4. Esteem needs are of two kinds. First is self-esteem, which results from competence or mastery of a task. Second, there is the attention from others and their recognition.
5. Self-actualization or self-fulfillment is the desire to reach the highest possible level of performance.

In a typical work environment, motivation concentrates on the higher two levels to bring the interests of each participating individual in alignment with the project goals through impersonal rules for rewards and punishment. The assumption of this *contractual* governance is that individualistic thinking and the objective calculations of self-interest are the main human motivators. A jelled team offer means to satisfy the social needs as well because it establishes channels for the free exchange, diffusion, appropriation, and integration of tacit and explicit knowledge. In this *connectual* governance, team members will view helping other members of the team succeed just like as helping themselves [Jin, 2001, p. 61–68]. Of course, this assumes that job security is provided.

Signs of a Jelled Team

Team formation goes through five stages denoted as forming, storming (start-up), norming (partial integration), performing (total integration), and adjourning (breaking up the team). When a team jells, its members will work collectively to explore news ways of looking at the problem, to assist each other in reaching their potential and to maintain a high standard of ethical behavior. Among the signs of a jelled team are [De Marco and Lister, 1987, p. 127]:

1. A strong sense of identity that satisfies the needs for belongingness and social integration and which is typically exhibited in the form of a low turn over
2. A feeling of joint ownership and an obvious enjoyment in doing the work that satisfies the needs for esteem
3. A sense of being in the *avant garde* and of doing something unique that satisfies the need for self-actualization

Effective project management entails spending the necessary time and effort to develop the team cohesion. The intensity of team building activities, however, should vary according to the type of innovation and the project phase. Contractual governance is useful in functional organizations and for incremental innovations, while the advantages of a connectual structure are apparent when tasks are highly interdependent—that is, when knowledge creation depends on the fusion of multiple disciplines, such as in architectural innovations. It is also needed when the transactional terms for knowledge creation and exchange are difficult to articulate in advance, such as in radical innovations. Furthermore, the rate of the jelling process depends on the cultural environment, taking longer when team members have to transcend their individualistic goals and agree to share a common vision. Thus, the recipes available on team building activities have to be tailored to the situation at hand.

It is worth noting that many start-up companies begin with an already "jelled team" because of the attractiveness of the concept that is being implemented or because the individuals involved had already worked together on other projects. This is a clear advantage that can help them in establishing themselves as credible players.

Enablers of Team Cohesiveness

Some of the factors that encourage team cohesiveness and binding are:

1. Respect—A disciplined approach promotes a professional atmosphere and inclusion of all team members encourages mutual respect.
2. Trust—Social scientists contend that people trust others in proportion to their self-esteem. One way to encourage trust is to stick to the rules established for the project.
3. Sense of purpose—A challenge can bring people together, provided that the situation is described in honest terms and that all team members can honestly talk about the problems they face so that remedial actions can be taken in time.
4. Care and mutual support.

5. Effective communication—This is the subject of the communication plan. Colocation is an important factor in building a projectized organization, particularly in the case of disruptive innovations.

Impediments to Team Consolidation

According to De Marco and Lister [1987, pp. 131–139], a *teamicide* strategy consists of procedures to make team formation very difficult or impossible. The elements of this strategy can be understood in light of Maslow's hierarchy of needs, because unmet needs at the lower levels of the hierarchy will frustrate attempts to satisfy needs at the higher level. The strategy comprises the following:

1. Robbing the team members of self-actualization
2. Robbing the team members of self-esteem
3. Robbing the team of belongingness
4. Robbing the team from the safety and security needs

No Self-Actualization. Pushing the delivery ahead to meet an arbitrary schedule with a lesser quality or a cheaper product undermines self-actualization. This will force team members to separate themselves as much as possible from the shoddy output that they are delivering as well as force the other team members to keep their self-image. With a quality plan, as explained in Chapter 8, it is possible to manage the project constraints without demoralizing the team. Such a plan will allow estimation of the delivery date using objective methods based on data describing the real situation or meeting business needs with a reduced set of features that operates satisfactorily.

No self-actualization is always a danger in incremental projects because the training is either on the job or for a very short time just to acquire rapidly some specific skills. In these cases, team members face the real danger of becoming technically stale; in fact, Deming's point no. 6 explicitly recommends developing people to improve quality [Deming, 1986].

In the telecommunications areas, people stay in projects because there are fascinating problems to be solved, because the pay is good, because their competence is respected and adequately rewarded, and because the work environment is good (e.g., free and honest communication, people are treated with respect, etc.). When people start thinking that they are just working for a paycheck at the end of the month—that is, that it is pointless to think of suggesting changes, because they will not be listened to or because those that expose themselves get fired—they usually detach themselves from their environment.

No Self-Esteem. Robbing the team members of their self-esteem happens when management shows that they don't trust their own people, particularly through excessive control. In a turbulent environment, executive leaders under high stress conditions often chose to restrict information, press for more conformity, and limit control to a few key individuals. Increased emphasis on efficiency and control often ends, thereby inhibiting the search for creative solutions and discouraging members from accepting any position of responsibility, because it is not clear that the hard work will be of any use. Quite often, the situation stimulates members to regain their self-esteem through power games that make no sense from a project perspective, but allows them to reclaim professional recognition or promotion, to deflect responsibilities, and to block policies that they don't approve of [Cardon, 1995].

No Belongingness. People working together need some time before they can feel comfortable with the operation of the team and discover how they can collaborate. Constant reorganization is an impediment; physical separation blocks the spontaneous flow of information. When people are geographically dispersed, the task of having a meeting through teleconference is an inconvenience because of the time difference. The pressure would be to reduce the number of such meetings. Thus, while the virtual teams are commonplace today, the core team involved in a disruptive innovation needs to be co-located.

Another way of impeding belongingness is to involve team members in many projects at once as a way to increase productivity. In fact, a study has shown that the short-term productivity increases with a second project assignment and then decreases when the same individual handles more than two projects at the same time. The increase in productivity with a second assignment comes from the ability of the team member to switch to the second project whenever the activities of the first project are delayed [Milosevic, 2003, pp. 202–203]. Such an increase may not be possible to sustain in the long-term unless there is an effort to acquire new knowledge and skills.

The absence of long-term perspective is translated in a high turnover, if the economic conditions are right, or inefficiency and low productivity. Other reasons are the feeling that individuals cannot make a difference, because decisions are concentrated in a few hands or because of lack of communication or recognition of contributions.

Turnover has several negative effects. First, the implementation gets delayed because of the lack of resources and the effort to look for a replacement. Next, time and effort are necessary to orient the replacement, which may be a burden on the rest of the team if the project is in full swing. Finally, the unexpected loss of the project manager—that is, a replacement that is not planned as indicated below—will have the greatest impact on the success of the project.

No Security. It can be easily understood that without job security, team formation is an elusive goal. In this case, the feeling that people can be replaced from one day to the next or that the job can be outsourced encourages them to spend most of the time cultivating their personal networks so that when the restructuring comes around, people will be able to find new jobs.

Team Breakup (Adjourning)

It is clear that jelled or self-directed teams just do not happen spontaneously, but that they must be carefully nurtured. Building such a team is costly in terms of effort and time. Unfortunately, many organizations do not recognize the value of such teams. For example, to cancel a project, there may be an intentional policy to disperse the members of a jelled team to overcome their resistance. Team breakup may also be an unintended consequence of a policy of gradual de-staffing a project as its activities wind up and assigning the available resource to any new project at hand without consideration to team preservation.

PROJECT LEADERSHIP

The project manager's style or approach to the project has an impact at all levels of the organization and all stakeholders—senior managers, project team members, functional managers, clients, and suppliers. Several researchers have studied defining the abilities and skills that successful project managers need to acquire and develop. We present the main results and show how they can be used in project management.

Transactional Versus Transformational Leadership

The literature on leadership effectiveness distinguishes between *transactional* or *exchange* leadership and *transformational* or *charismatic* leadership. A transactional leader focuses on the tactical issues of daily operations within the current constraints. The leadership is mostly concerned with efficient procedural, managerial, and operational functions with the aim of improving the coordination of activities and the optimum allocation of resources to resolve task-related complexities. A system of rewards and punishment is the medium of exchange between the leader and the followers. A transactional leadership is most suited to hierarchical authority structures, or when there are standardized procedures detailing what to do and what to avoid, thereby reducing the variability and allowing a focused concentration on efficiency.

A transformational leader articulates a compelling vision that mobilizes the followers and engages them emotionally and intellectually to surpass themselves in the call of duty. One characteristic of transformational leaders is that they succeed in changing the values of their followers. In an organization, the charismatic appears through interpersonal binding with respect, trust, and loyalty [Limerick and Cunnington, 1993, pp. 140, 180–190, 205–208]. Consequently, a transformational style is better suited to cases where the reward is intrinsic, the authority is diffuse, and decision-making is decentralized [Howell, 1997].

Project Manager's Authority. Interdependence is fundamentally a relationship of power and influence, because a leader cannot lead without followers. In a power relationship, one party is able to control and to guide the process and secure the desired outcome. Influence is a strategy where a party achieves its goal without explicitly using its power, or, in the words of Sun Tzu, "to subdue the enemy without fighting is the acme of skill" [Tzu, 1963, p. 77].

The project manager role is operational in nature. *Formal authority* (sometimes called *legitimate power*) derives from the hierarchical position of the individual within the organization and the authority to hire and fire, to control budgets, or to review and appraise activities [Knutson and Bitz, 1991, pp. 26–31]. In general, people accept orders from legitimate power figures, regardless of ethics or morality, as has been demonstrated by the famous experiment where subjects were asked to increase the intensity of (fake) electric shocks to those who would give incorrect answers, even to the point of causing pain. Informal authority comes from experience/knowledge, association with other powerful people, credibility and integrity, and so on. This informal authority of project managers is essential for their ability to integrate all the efforts expended on the project and to negotiate with the functional managers to give direction on what needs to be done, when it is needed, and how much effort should be expended.

Manipulative Behavior. There is a fine line between leading people and manipulating them: Leadership is sometimes defined as the ability to get others to commit to doing things they would not have otherwise done. A manipulative leader will act in a deceitful way through disinformation and information retention. Some of the signs of manipulation are:

- Values conflicts are settled by coercion by the leader.
- Leaders do not live up to the high standards they set for everyone else.
- Policies, procedures, and processes do not promote ethical behavior.
- Leaders stretch the facts to engage in impression management.
- The leader encourages the follower's dependency instead of fostering their development as future leaders by allowing them to think for themselves.
- Benefits are not equally shared or are not proportional to the sacrifice; instead, they are reserved to the leaders and their entourage.

Typical examples for impression management include misidentifying threats and opportunities as well as phony deadlines. To elicit a rapid and radical organizational response to environmental changes, leaders may label them as "threats" to consolidate power and suppress dissent. Or they may generate an unfounded enthusiasm by describing the situation as an "opportunity" even though loss is very likely and the group has limited control over the situation. Thus describing a challenge as a "threat" or an "opportunity" is more than just a linguistic precision but can be a deceptive practice [Howell, 1997].

Unrealistic or phony milestones usually backfire and demoralize people. In the typical phony deadline scheme, the deadline is imposed even though it is evidently impossible to meet and everyone knows it but is not allowed to say so: The scope of the project is not changed, the performance requirements remains the same, and the available resources are limited. Eventually, the deadline is missed and a new deadline is set, a sign that the management does not know what they are asking or that they do not trust the team's judgment.

The drawback of such tactics is that group members can see through them eventually. Once they do, it is extremely difficult for the management to regain their trust.

MBTI Classification of Leadership Styles

Leadership styles can be categorized using the MBTI criteria into four profiles: the explorer, the coordinator, the driver, and the administrator [Myers et al., 1998; Wideman, 1998; Wideman and Shenhar, 2001].

The *explorer* or the *entrepreneur* is someone looking for new opportunities and ways to change the status quo. People in this category are imaginative, restless, and on the lookout for opportunities and improvements. They are comfortable in the lead and exude confidence and are good at networking and selling, but they are less attentive to the day-to-day details.

The *coordinator* exhibits the ability to organize of work, particularly in conflict situations. Coordinators act as facilitators, listening to team members to synthesize the various views in a consensus that they can all relate to. Even though they will delay making a decision until such a consensus arises, they are capable of taking a more independent viewpoint and detaching themselves from the environment.

The *driver* type is associated with action-oriented persons that are hard-working and hard-driving and can exercise self-discipline. They plan well and can make decisions under pressure or with incomplete information by concentrating on their goals. In addition, they are pragmatic, realistic, resourceful, resolute, and hard-driving. Another characteristic is their perseverance and the ability to carry the ideas through as well as to delegate responsibilities.

The *administrator* type comprises those that recognize the need for stability to optimize productivity. They are good at gathering and evaluating information carefully, with thought given to anticipating conflicts.

The correspondence of the project leadership profiles with the MBTI types for the U.S. population is shown in Figure 7.4.

This classification correlates with the results from another study focused on the products that won awards for design excellence at the annual Hannover Fair, in Germany, in 1993 and 1994 [Hauschildt et al., 2000]. It turned out that the capabilities of project managers can be described using seven factors: ability to operate under stress, experience, decision-making process, creativity and idea generation, interpersonal communication, ability to motivate others, and integrative thinking. Using these factors, the project managers of the winning products could be clustered in five categories [Hauschildt et al., 2000]:

1. Project stars that have above-average scores on all capabilities (20% of the sample).
2. Promising newcomers who show above-average capability in all areas except in organizing under conflict and are inexperienced (20% of the sample).

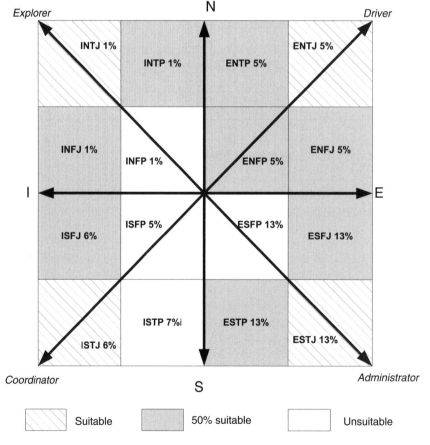

Figure 7.4 Relation between MBTI grid, project profiles, and suitability for project work in the U.S. population.

3. Focused creative experts who can carry their own ideas but are not very good at collective activities such as cooperative organization and integrated thinking (11% of the sample).

4. Uncreative decision-makers who depend on others for new ideas but are capable of making decisions (11% of the sample).

5. The thick-skinned pragmatists who are capable of handling conflicts even though their capabilities are average (38% of the sample).

Clearly, studies in different contexts have shown that some kinds of project managers are easier to find than others and that the kind of project manager should be matched to the type of projects with which they are most likely to succeed.

Finally, Figure 7.5 maps the percentage of personality types of top managers in the United States, taken from Kroeger and Thuesen [1992] as cited in Krumwiede and Lavelle [2000], over the MBTI grid. Table 7.4 contains the top manager preferences. This shows the weight of Thinking (T) and Judging (J) preferences among top U.S. managers. Interestingly, those top managers with a strong Intuiting (N) preference are statistically associated with organizations that have embraced total quality management (TQM) [Krumwiede and Lavelle, 2000]. A possible explanation is that those with the N preference are more likely to plan for the long term and to be patient enough to wait for results.

Time-Dependent Leadership

It is well understood that different stages of a company's life require different approaches [Witzel, 2004] and that, likewise, projects in different periods of their life cycles need different approaches and styles [Wideman and Shenhar, 2001]. In other words, there is no single profile that is valid all the time in a project life cycle. The change of leadership style often calls for changing the leaders themselves, a process that is fraught with dangers and may be destructive if the leadership does not accept the need for change and resist the pressures to change the mode of operation. For example, when a shift from the transformational to the transactional style is warranted, because of the intense emotional involvement and the individualized relationship that characterizes the bonds among the group, those who disagree with the leaders may be "branded as heretics and subjected to social ostracization, threats and punishments" [Howell, 1997]. On the other hand, transactional leaders under stress due to disruptive changes or environmental threats can resort to more centralization of decision-making and control, increasing pressures to conform to the hierarchy, and restricting the control of information. The antidote to this may be a prior agreement among all stakeholders on what profiles are suited for each phase as a function of the degree of maturity of the technology, of the type of innovation, and of the project cycle.

Table 7.4 Distribution of Preferences of Top Managers Versus U.S. Population

Preference	Population	Top Managers	Opposite Preference	Population	Top Managers
Sensing (S)	76%	66%	Intuiting (N)	24%	34%
Thinking (T)	51%	95%	Feeling (F)	49%	5%
Judging (J)	50%	88%	Perceiving (P)	50%	12%
Extroverted (E)	72%	47%	Introverted (I)	28%	53%

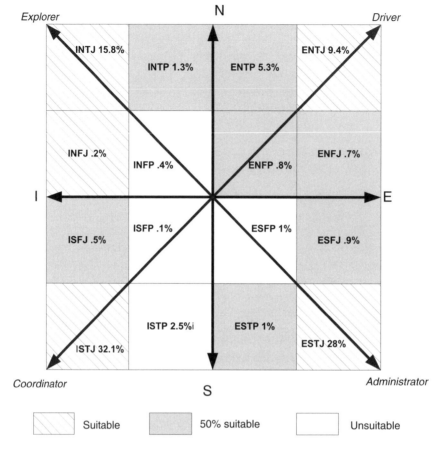

Figure 7.5   The MBTI grid, project manager profiles, and personality type of top managers in the United States (the percentages showing the approximate proportion managers of that type).

Matching Leadership Style with the Project Phase. In Chapter 3, the life cycle of a typical telecommunications service project was described to have five major phases: concept definition, initiation and preliminary planning, implementation, controlled introduction, general availability, and closeout. The explorer style of leadership is appropriate during the concept period, while the coordinator type is suitable for the initiation and preliminary planning period. The driver, because of its assertiveness, is suitable during the implementation and the controlled introduction. Finally, the administrator type is most helpful during the closeout and transition to life-cycle operations.

Another way of looking at project leadership is by considering the characteristics of the work group as it changes throughout the project and how these changes affect the role of the project manager [Thamhain, 2001, 2003]. In this approach, when the team is being formed—that is, in the initiation and planning phase of the project—a more directive style of management is needed to give clear directions. Later, as the project progresses, leaders must pay strong attention to the human side, build confidence in the team capability, and deal with issues of workload, performance measures, and integration.

Matching Leadership Style with Innovation Type. A transformational leadership is useful for both radical and architectural innovations. In these two categories, the exploration needs persons able to take risks, act under uncertainty [Hofstede, 1980, 1990, 1997], and look into the future persons conditioned not to avoid uncertainty. The classifications on the MTBI grid provide another look at this factor. The explorer is most suitable for radical innovations, while the coordinator has the profile needed for architectural innovations [Wideman, 1998]. Incremental innovations depend on the skills of the administrator. The profile that is most suitable for platform innovations is that of the driver.

Matching Leadership with Technology Maturity. At the start-up phase, the characteristics of the technology, product, or service are not well-defined. Because the search for innovative products or services is underway, the production process is not stable and knowledge is obtained through trials and error. The emphasis is on problem identification and solving calls for a learning organization with a structure supportive of experimentation and cooperation. The challenges facing a learning organization are not restricted to the technical aspects of the job but are, instead, restricted to how to capitalize on the mistakes for future growth: mistakes in selecting the vendor, in establishing internal communication channels taking into account various diversities, or in scheduling the various tasks, and so on. Furthermore, the fluidity of organizational, technical, and environmental conditions requires the project manager to possess sufficient prestige (for example, through technical breadth) to earn the respect of the team. Project managers must be able to sense opportunities, define appropriate goals, and articulate them in a compelling manner. They must also build trust among the team members to share and document all what is being learned for future use. They must have political skills to negotiate the characteristics of the evolving boundaries of the project organization with the environment. These are the characteristics of transformational leadership.

Projects of the growth phase are platform innovations to expand the customer base, consolidate the market share, and support a scalable service offering. Flexible planning increases adaptability to changing market conditions. The organization is typically networked (in a matrix form), exploiting technical and marketing expertise to identify additional market segments where the product or service could be adapted. In addition to the technical aspects, the project manager works with the sales team and negotiates with the vendor a set of deliverables and pay points. Thus, successful project managers should be able to delegate—that is, leave the technical details to the technical experts—and focus on growth-related capabilities. The project management will have to make quick decisions with limited or partial information, handle conflicting requirements, and deal with different types of people.

Finally, in the maturity phase, there is a period of continuous improvements with the objective of reducing the production and distribution costs through increased efficiency. The role of the project manager is to find the optimal combination of formal rules and procedures to perform the project tasks through work segmentation, allocation of specialized labor, hierarchical decision-making, and so on. This is when people can focus more on delivering their own targets. In incremental projects, technical ability is critical for gaining credibility, thereby obtaining influence with members of the project team. In small companies, in particular, the technical competencies of the project manager—coupled with assistance from the rest of the organization, such as the marketing and product management organization—may cover the lack of basic skills such as scheduling and budgeting.

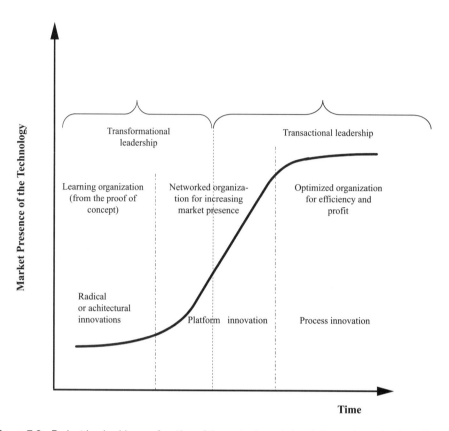

Figure 7.6   Project leadership as a function of the project needs in relation to the technology life cy-cle and the type of innovation.

Figure 7.6 illustrates the relation of the leadership style and the S-curve of technology maturity and the various innovations types.

CONFLICT RESOLUTION

The project progress relies on synchronized execution of the project tasks and information sharing among stakeholders, including sponsors, suppliers, and customers. While some diversity—both professional and personal—is desirable, people in a team have to be able to relate and collaborate with each other. Differences in profiles among people, along with the stresses of the project execution and synchronization, provoke tensions that may lead to either (a) divisiveness or (b) improved opportunities of additional learning.

The origins of the conflicts can be ascribed either to the contractual structure of the project or to the connectual structure that link all the stakeholders.

Conflicts Due to Contractual Structures

In 1959, Frederick Herzberg developed a list of factors that affect motivation. Hygiene factors are turn-offs related to the contractual relations within the work environment.

These negative factors must be neutralized before motivators can be effective. Motivators are those that encourage workers and are related to Maslow's hierarchy (see Table 7.5).

The origin of many hygiene factors reside in the contractual structure and its processes that constrain the way the group operates and interacts with the environment. Some of the process conflicts could arise from the dual reporting structure that is typical of matrix organizations, from upper management interventions to undercut the project manager authority, or from the unavailability of some key members of the project team. For example, the intervention of an executive may force missing some milestone, or some unit may not be able to provide the needed support within the allotted time. Fixing process problems call for a redrawing of the contractual relations, a task that exceeds the authority of the project manager, unless the organization is of the projectized format. Group interdependence increases the negative effects of process conflicts on the short-term performance of the project team.

Conflicts Due to Connectual Structures

To enhance collaborative relationships, the sources of conflicts need to be taken into account to build collaborative relationships. However, the forms that these tensions take may be culturally dependent; therefore, it is important to learn how to read and interpret the cultural cues before shaping a response. We discuss now the types of diversity and how they can be related to conflicts within the project team.

Types of Diversity. Recent research has shown that diversity can be social, informational, or value-related [Jehn, 1999; Jehn et al., 1999; Neal et al., 1999]. *Social diversity* is a function of who does the task, and it appears in the form of explicit differences that are directly observable such as physical characteristics (race, age, and gender) or responses to objective conditions such as noise, crowdedness, stress, and so on. Conflicts around social diversity can be related to personal, technical, or environmental preferences (e.g., like working under pressure) and enjoy tight schedules and high-pressure situations while others contribute better under a relaxed environment.

Informational diversity relates to technical background, education, and prior work experience. Conflicts due to functional diversity arise in cross-functional teams; team tenure diversity is typical in projects of longer duration, because some members rotate on and off the project, while others remain throughout. New members may also bring a fresh perspective that could challenge accepted assumptions. Associated disagreements and contents relate to the content of tasks—that is, what to do, what norms, standards or best practices to follow, and so on. These exchanges are useful when the task is complex or re-

Table 7.5 Herzberg's Factors for Hygiene Versus Motivation

Hygiene or Dissatisfiers	Motivators or Satisfiers
Company policies and administrative practices	Achievement
Supervision	Recognition
Relationship with boss	Work itself
Work conditions	Responsibility
Salary	Advancement
Relationship with peers	Growth

quires specialized knowledge or if the degree of uncertainty is high (the problem is new and there are few defined procedures).

Value diversity refers to strategy differences on how, why, and when the task should be done. Value conflicts relate to disagreements on the group's mission, on the distribution of power and resources, and on the assignment of responsibilities. Values are at play in the identification of ethical dilemma or problems. Value diversity appears when decisions have to be made with limited information because such choices are never free from risk. The nature of assumed risks depends on the perspective and whether a person is optimistic or pessimistic, accepts or avoids risks, and so on. Typically, value divergences appear if the scope of the project was not well-defined or well-controlled or if shifts in the environment have made the current organization inappropriate.

Congruence of values among a group's member takes some socialization into the new environment. Successive reorganizations prevents this congruence, which could lead to value conflicts.

Figure 7.7 illustrates how the various types of diversity appear in the form of personal, tactical, and strategic differences, and it also illustrates the mutual influences of each type of diversity.

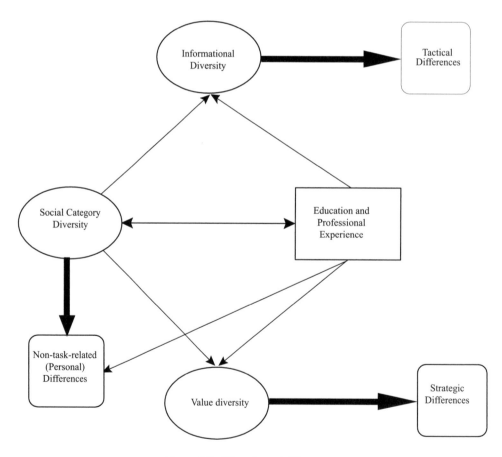

Figure 7.7   Diversity and differences.

Examples of Social Diversity. Conflicts related to social diversity can be imputed to communication styles such as silence, brevity, or abruptness, which can be considered as an insult, a put-down, a blame, a threat, and so on. Conflicts due to social diversity are also reflected in exclusionary behaviors in organizations.

Examples of Informational Diversity. Compared to those with a background in humanities, the typical attitude of engineers and scientists is characterized by several traits such as (1) a binary orientation towards issues (right/wrong, black/white), (2) micromanagement and focusing on the details instead of the big picture, (3) respect for technical proofs, and (4) less attention to money and power [Tingstad, 1991, pp. 3–5].

A study of time-to-market of new product development projects in a high-technology company found that project engineers working with engineers from other divisions took 20–30% longer to complete their projects when they had not established a personal relationship. This is why when team members come from different functional backgrounds, some time must be allotted for them to establish a common vocabulary and learn to trust each other's judgments [Hansen, 2002].

Another example of information diversity is given by the subtle but real differences between the fundamental assumptions of telecommunications and information technology designs. In telecommunications, which is based on audio and visual real-time exchanges, the focus is on making remote interactivity as natural as possible—that is, on the compatibility of signaling protocols and electrical interfaces and on the reliability and full-time availability of the networking infrastructure. The designers of intelligent terminals, in contrast, build machines to extend human capabilities for storing, interpreting, and comprehending information. The market for intelligent end-user terminals is characterized by polychronic activities ("multitasking"), short product life cycles, intense rivalry, constantly emerging substitutes, and the need for differentiation. In contrast, the world of public telecommunications relies on long-term planning, cooperation, orderly integration, and homogeneity. As a consequence, the assumptions about time and space in the two fields are also different. For example, the average life cycle of an elegant "software solution" for the end-user is 18 months, which is merely the time that the initial layout of wide area networks.

Another difference appears in the field of standardization to ensure end-to-end compatibility. While voice communications need standards to ensure similarity between the two endpoints, decentralized network management may be sufficient for data services and standards are often needed merely to ensure the compatibility of the equipment. The convergence of the telecommunications and information underlines the contrast between more conservative careful procedures and aggressive approaches to ensure reliability, availability, and predictability of interactive traffic.

Examples of Value Diversity. Telecommunication projects today are increasingly complex and assemble teams from a variety of companies and many countries: undersea cable systems, Olympic games, outsourced projects, and so on. In these and similar situations, values that are taken for granted in a culturally homogeneous environment are now in question. In light of the pioneering works by Hall [1959, 1976, 1990] and Hofstede [1980, 1990, 1997], project managers must take into account the style of authority and allocation of power (centralized versus decentralized, consensus versus authoritarian), the attitude toward time, and various work practices [Sherif, 2000].

Attitudes Toward Authority and Rules. Whether a decision is taken in a top-down or a bottom-up fashion, in a centralized or decentralized fashion, is a cultural phenomenon. Similarly, adherence to procedures varies from culture to culture, as does the strictness with which these procedures should be followed.

Hofstede's concept of "power distance" provides a rough assessment of the situation by measuring the ability of employees to decide independently of their hierarchical superiors. When used with caution, this tool may help in deciding whether the leader should be directive or participative. The caution is warranted because decision styles that are close on the "power distance" scale do not necessarily correspond to the same daily practices.

The Concept of Time. Time is one of the constraints of projects, yet cultural difference in their understanding and approach to this constraint can be a cause of conflicts and misunderstandings. For example, U.S. project managers like quicker decisions than European project managers, are more inclined to allow groups to determine vacation and work schedules, and use subordinate input more frequently. European project managers, however, prefer group decisions [Lee et al., 1995]. It should be noted, however, that the "European" managers differ significantly among themselves [Breuer and de Bartha, 1993].

The cost–benefit analysis of long-term plans raises issues concerning the distributions of benefits and risks; accommodating those concerns is a time-consuming process. If the focus is on the short term, the purpose of the activities is to grab available opportunities using existing information without the benefit of a thorough analysis. The goal would be to learn on the job to do the task fast rather than to learn well. This focus on the present is very useful in a start-up mode in the case of radical or architectural innovation because knowledge is being built up. However, in this mode of operation, each individual is dependent on the group support and assistance to supplement the lack of information. Clearly, differences with respect to time appear in the way planning aims at obsolescence or at resilience.

With respect to the attitude to time during task execution, cultures can be divided into monochronic or polychronic [Hall, 1959; Hall and Hall, 1990]. Polychronic people can perform many tasks simultaneously, while monochronic individuals typically envision time in a linear and compartmentalized way. In polychronic cultures, a vast and elaborate information network of clients, friends, and family must be maintained scrupulously so that they remain effective.

Work Practices. Culture values are the foundation for a collective, shared vision. Every culture has to balance issues of individualism and communitarism. According to Hofstede [1980, 1990, 1997], cultural differences affect the working atmosphere and whether assertiveness or competitiveness are valued at the expense of care and attention. They also affect the organization of space and the way collaborative work takes place. In a high-context society, the decision-maker must be at the center of the information flow. Whether admitting faults or failures is a strength or weakness, or whether the participants need to be knowledgeable on all aspects associated with their status, are cultural variations [d'Iribarne, 1989; Winch et al., 2000].

Conflicts and Diversity. Whether conflicts due to the connectual structure should be suppressed or can be managed to benefit the project depends on its root causes. In general, informational diversity leads to improved team performance, because they stimulate a search for new facts and encourage experimentation. The impact of task-related conflicts on the group morale depends on whether the dominant culture accepts contradictory debates as a legitimate way to seek truth. Social conflicts can occupy the team and lead to divisiveness or to poorer quality decisions. The resolution of value conflicts need some compromise on all sides because imposing a solution means that the team members will be less committed to these decisions: As soon as the project manager turns his or her back, things will revert to their old ways.

Effects of Conflict on Project Performance

Differences can have positive and or negative effects on the performance of a team. Figure 7.8 summarizes the results [Cardon, 1995; Jehn, 1999; Jehn et al., 1999; Holahan and Mooney, 2004; Neal et al., 1999]. Constructive conflicts occur when team members disagree on the specifics of the task at hand. These debates improve the decision-making process because they help team members in getting a better grasp of the underlying issues and work to synthesize multiple approaches to produce a solution that is superior to the original proposals. The effects of task conflicts on morale depend on the situation of the team, which means that there are other effects (such as cultural considerations) that enter into play. This is why informational diversity is more likely to increase workgroup performance when the group members share the same values.

Destructive conflicts occur when team members engage in power games to corner each

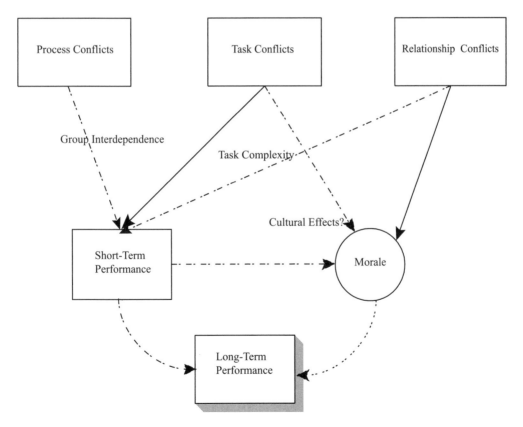

Figure 7.8 Effects of conflicts on team performance.

other, creating tension and animosity, planting distrust among the team members, and draining collective energies from the tasks to be accomplished. The effect of relationship conflicts on short-term performance is compounded when the tasks are complex. There are many undeclared reasons for such conflicts: diverting attention from the real problems and responsibilities, attracting attention to the protagonists, embarrassing opponents, and so on [Cardon, 1995]. True value conflicts seem to occur when (1) there are significant value different among the stakeholders, (2) there are credible alternatives from a legal and ethical viewpoint, and (3) the benefits and risks of each of these alternatives are not equally distributed; that is, for each choice the winners and the losers are not the same and/or the gains or losses are not shared uniformly. Groups that have greater value diversity tend to be less effective and efficient and have a poorer performance and lower morale. The relationship between short-term and long-term performance is not always clear.

Dealing with Conflicts

Typical techniques for dealing with conflicts are

1. Problem solving
2. Coercion
3. Compromise
4. Accommodation
5. Withdrawal or avoidance

Problem Solving. Problem solving is a creative approach in which the participants work together to define a win–win solution. Activities include steps to bound the problem, go the real issues, collect information, develop alternatives, and propose a solution that removes the root cause. The basic principle is to dissociate the symptoms (i.e., the people) from the system and reconfigure the whole process to allow each party to get their needs.

Coercion. Coercion is the imposition of a solution by the dominant party without trying to resolve the underlying issues. Coercion may be needed to impose ethical norms or quality standards or to resolve escalations, when peer groups are unable to resolve their conflicts due contradictory demands on resources. Forcing may be also needed in times of emergencies, provided those who have the legitimate authority endorse it. It is useful in the early stages of team formation to give a clear direction. However, such a style is counterproductive for the long term, and decisions made with this style may need to be revisited when the urgency disappears or the situation stabilizes. As a general policy for solving differences, coercion is typically an expensive strategy in terms of (a) the cost of implementation and (b) the cohesion of the group.

Compromise. A compromise is the typical approach when all parties agree on the principles yet differ on the details, when all have to win but need to maintain a working relationship with the other stakeholders. The outcome is something that no party is totally happy with but still can accept. This can be compared to price negotiation between a buyer and a seller. In a project, a compromise is often to balance personal needs and project priorities. This is also a common method for defining telecommunication standards by including options or different profiles for different applications.

Accommodation. This is a position to emphasize the common objectives and deemphasize the differences to create good will, to save time, to alleviate tensions, and to keep the appearance of friendliness. Accommodation is a way to handle valued conflicts that do not affect the project work. For example, in preparing meeting agendas, punctual people could be first scheduled to allow tardy people the time to attend without delaying all the other attendees. When the team is highly united and committed to achieve the established project objectives, some of the team's idiosyncrasies may be accommodated by delegating to the team members local decisions for the team to operate in a "self-directed" fashion.

Withdrawal or Avoidance. Avoidance of confrontation is the safest solution when the stakes are low and/or there is not a possibility of removing the root cause. For example, the conflicts may be only the public face of hidden agendas or attempts to divert attention from the real issues [Cardon, 1995].

SUMMARY

Capable contributors are indispensable for the success of any endeavor; furthermore, the cohesion of the project team contributes to the smooth integration of individual efforts. Effective teamwork is a critical determinant of project success and the organization's ability to learn from its experiences and position itself for future growth. The style of project permeates the environment of project execution and defines the tactical direction of the project and their style of operation. Individuals with the required traits for project work are not evenly distributed in the population, and not everybody is comfortable working in organizations with a matrix reporting structure or which are totally project-oriented. Successful human resource management depends also on understanding the relationship between the organizational dynamics and the characteristics of the work to be executed such as the phase of the project, the innovation type, and the life cycle of the main technology involved. Strategic differences in a project may be due to value conflicts, especially when there is a radical shift in the environment that make the current project structure inadapted to the work. Diversity can be extremely useful if conflicts can be managed to improve the service implementation or improve the project organization. Effective conflict management requires that the appropriate method be used at the right time.

8

QUALITY MANAGEMENT

OVERVIEW

In equipment design and development, management of the project quality is intrinsically different from the management of ongoing operations on the factory floor. In contrast, in telecommunications services, the service itself is the product [Ward, 1998, p. 35] and the boundary between "project quality" and "operations quality" is not as sharp. In the literature, quality assurance describes the preventive action of regularly evaluating the overall project performance to provide confidence that the relevant quality standards will be met. The scope of quality control is to verify that the project deliverables comply with the quality standards set and, if needed, to take the appropriate corrective actions to remove the causes of unsatisfactory results. Such a distinction is difficult to make in service development. For example, if a service is difficult to set up—perhaps due to shortcuts made to on implementation costs—installation could be longer than promised or susceptible to operator's errors. In reality, there are two sets of overlapping quality requirements for telecommunication services, one associated with network operation and maintenance and the other related to the end-user's experience. These internal users include the on-site workforce that maintains the daily operation from various network operations centers as well as those involved in sales and provisioning of the service. Satisfaction of both sets of users/customers is equally important.

Of course a project well executed and conducted by following the best practices of project management may still deliver a disastrous product if the original concept is flawed or if the day-to-day running of the service is deficient. However, this is outside of the responsibility of the project team.

Managing Projects in Telecommunication Services. By Mostafa Hashem Sherif
Copyright © 2006 The Institute of Electrical and Electronics Engineers, Inc.

QUALITY AND INNOVATION

In theory, quality planning begins with the identification of the list of users or customers; the next step is to discover their needs and translate them into project specifications using tools such as the house of quality [Kloppengorg and Petrick, 2002, pp. 28–29]. This process is relatively straightforward in the case of sustaining innovations. Tacit or explicit knowledge for the past are applicable and, if the organization is stable enough, oral transmission of this knowledge to newer members of the team is possible through social interactions. In incremental service innovations, which are typically process improvements to reduce cost or improve the competitive position of the network operator, the conduct of the project itself can be evaluated on the basis of conformance to budget, schedule and scope. Quality planning for service platform innovations is more involved because they incorporate a quantum leap in technology, which increases the uncertainties on the project's budget and schedule. Nevertheless, quality criteria remain mostly defined from a technical perspective.

The case of disruptive innovations is considerably more complex. While the internal customers and their needs remain more or less the same: easy to use systems, conformity with existing with work procedures, etc., the profile of the external customers remains fuzzy and their expectations tentative. Past lessons are not necessarily applicable and in some cases the "wheel" may have to be reinvented. Finally, alignment of the project with organizational objectives is trickier particularly in times of rapid change. As a consequence, decisions on quality pass from the exclusive hands of engineers to many others including lawyers and accountants. To understand why this is so, we take advantage of the concept of the cost of quality due to P. Crosby [1979, pp. 124–125].

Accordingly, we distinguish among three types of cost in for quality in telecommunication service projects:

- *Prevention cost* is the cost of all activities undertaken to prevent known defects from affecting the service from meeting the levels agreed to. This is the cost associated with going around defects that are discovered while the product is in service or that could not be fixed before the service was generally available.

- *Appraisal cost* is the cost incurred in the evaluation of the equipment that support the service due to inspections, tests, vendor management, and so on. This item includes the salaries for the quality team members, cost of hardware used for testing, cost of software licenses, and so on, as well as the opportunity cost for delaying the service introduction.

- *Failure cost* is the cost of experiencing a failure during operation. It covers the penalties as well as the cost of failure containment and recovery.

The project sponsor usually sets the upper bound on the total cost of quality, even though it is not always explicitly stated. Let us now consider some examples of the cost of service failures.

Quality of telephony over the public switched telephone network (PSTN) is tightly regulated. In the United States, for example, the Federal Communications Commission (FCC) stipulates that in public telephone networks, a major outage is a service failure that either (a) lasts 30 or more minutes and disrupts the service to more than 1000 subscribers or (b) causes total loss of service to a government emergency response agency. Furthermore, the FCC requires that network operators report to its Network Reliability Council

(NRC) any telephone service outage that affects 30,000 subscribers for over 30 minutes. In addition, the Commission requires the reporting of outages affecting "special offices and facilities," such as major airports, 911 facilities, nuclear power plants, major military installations, and key government facilities, regardless of the number of customers affected (Report No. DC-2626, July 14, 1994, CC DOCKET 91-273). Some states have more stringent requirements; for example, the state of Washington has reduced the number of affected customers to 1000 subscribers.

The quality of the service for "enhanced services" (i.e., data networks or voice over IP networks) is left to the agreements between customers and their service providers as defined by service level agreements (SLA). Typically, a major outage is an event that either (i) results in more than 5% of the active circuits (or ports) to become unusable for 30 minutes or longer or (ii) results in more than 10% of the active ports to become unusable regardless of the duration. The consequences of such an outage depend on customer's applications. Table 8.1 gives estimated losses when telecommunications failures affect mission-critical applications in selected business categories [Pesola, 2004]. Depending on the SLA, operators may be obliged to pay some compensation, in addition to potentially embarrassing publicity. Table 8.2 gives an example of a refund policy due to outages. In addition, in the event that the service fails to adhere to the service levels for a period of, say, 3 months, and the corrective measures introduced by the service provider does not improve the performance, the customer may terminate its agreements.

Service Release Management

Because of the above considerations, service release management is risk averse. Project management contributes to the decision process by providing a systematic method to balance the various components of the cost of quality and to evaluate whether further delay in service introduction because of additional testing can be justified in light of the penalties of finding defects in the field. The tradeoff is described with the following inequality:

$$\text{Total cost of quality} = \text{Prevention cost} + \text{appraisal cost} + \text{failure cost}$$

$$< \text{Upper bound for the cost of quality.} \qquad (1)$$

Project activities associated with appraisal (e.g., testing the equipment to verify that they are stable and perform at the level that was contracted for) should continue as long as the appraisal cost is less than the cost of prevention and the failure cost, that is,

$$\text{Appraisal cost} < \text{Prevention cost} + \text{failure cost.} \qquad (2)$$

Table 8.1 Estimated Business Losses if Mission-Critical Applications Fail

Type of Business	Estimated Loss in U.S. K$ per Minute
Brokerage	100–200
Manufacturing	50–100
Point-of-Sale	20–100
Travel Agency	1–10

Table 8.2   Example of a Refund Policy in a Service Level Agreement (SLA)

Duration of Outage	Refund as Percent of Annual Rent
> 2 hours and < 3 hours	2%
> 3 hours and < 4 hours	4%
> 4 hours and < 6 hours	6%
> 6 hours on any one occasion	8.3% (1 month of free rental)

Equation (2) compares the cost of defect discovery during service with the cost of delaying the deployment to continue testing. The strategy that these two inequalities indicate allows engineers to fulfill their two-sided responsibilities of contributing to the bottom line of their employers while keeping their technical integrity. It is also consistent with Deming's point No. 11 [Deming, 1986] that *there is no point in defining numerical standards and goals to improve the situation before understanding the root cause of the special and common cause variations.* Of course, an objective method to estimate the expected cost of failures is still needed to avoid arbitrary goals or slogans.

QUALITY PLAN

The basic principle of quality management engineering is to strive from continuous improvement based on facts by removing guesswork and personal biases as much as possible. Therefore, the quality plan must specify the following:

1. The quality targets, which are performance limits within which there is confidence that the service being developed will be able to meet the sponsor's objectives. Definition of the quality targets beforehand would prevent future disagreements on the level of quality needed.
2. The resources allocated for implementation, including the features to be tested, the methods for testing and the expected behavior. In this way, the plan documents all the knowledge that was available at the time the tests were conducted.
3. The data elements to be gathered, compiled, and interpreted as well as the data acquisition and processing system. The accuracy of data collection affects the pertinence of the reliability analysis. For example, hardware failures should not be included in the analysis of software reliability, and decisions have to consider all those features that are slated for deployment.
4. The method of data analysis and the project metrics to track the project progress, to verify that the targets are met, and to determine if the service is ready for deployment and the associated risks.
5. The improvement plan to close the gaps between the observed and the desired behaviors. This usually includes some communication with the vendor.
6. The communication system to inform all project participants of the status of the quality evaluation.
7. Methods to store, retrieve, and preserve for the future the lessons learned during the evaluation.

In incremental innovations, the quality plan can be quite detailed. In other innovations, however, the quality plan is built through successive iterations as the quality team improves its expertise. The first round of testing focuses on the functional characteristics of the service and is followed by as many additional rounds as needed to meet the quality objectives. Maintenance release testing continues until all critical problems and most major defects that have been discovered are fixed. The number of expected remaining failures should be within the quality objectives defined in the statement of work.

Categorization of the Defects: Urgency and Criticality

Not all defects have the same impact on the network operation or on customers' traffic. The categorization of the defects can be achieved using a two-dimensional grid to reflect the severity of the defect and the priority of the fix—that is, the criticality of the problem and the urgency of the fix. The repair of a highly severe problem may be postponed, if its effects are confined are to areas outside the perimeter of the service to be developed. Figure 8.1 depicts this two-dimension classification with three levels per dimension.

Foremost attention should be on defects that fall in the upper left-hand corner—that is, those that have the highest combined effect. Priority is a strictly business decision that depends on the expectation of the project sponsor, particularly in projects with phased deployment. With respect to the criticality, there are currently two main classification schemes for the severity of anomalies: The first comes from Telcordia Technologies and enjoys the support of operators of public networks, while the second comes from the Institute of Electric and Electronic Engineers (IEEE), is common in private (enterprise) networks.

Table 8.3 summarizes the Telcordia criteria for assigning defect severity as defined in GR-929-CORE, Appendix A [Telcordia, 2002]. The classification of IEEE Standard 1044.1-1995 with five degrees of severity is shown in Table 8.4.

According to the Telcordia classification, *critical* (or severity 1) defects are associated with functional or stress-related malfunctions, which, if deployed, will disrupt the services offered to customers or will block further testing. There is no acceptable workaround. *Major* (or severity 2) defects cause a degradation of service or disrupt the opera-

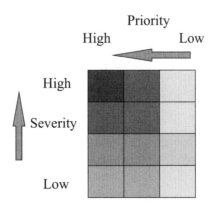

Figure 8.1 Classification of defects in terms of severity and priority.

Table 8.3 Description of Severity Levels for Defect Classification

Level of Severity	Status of service	Description
Critical	Failed	Problems that severely affect service, traffic, billing, and maintenance capabilities, and require immediate corrective action, regardless of time of day or day of the week.
Major	Degraded	Problems that cause conditions that seriously affect system operation, maintenance, and administration, etc., and require immediate attention. The urgency is less than in critical situations because of a lesser immediate or impending effect on system performance, customers, and operation and review, perhaps because of an acceptable workaround.
Minor	Acceptable	Problems that do not significantly impair the functioning of the system and do not significantly affect service to customers. They may not associated with the planned services or may be tolerable during system use.

tions of the network operator. Although without an intervention of the operations support personnel they impact the quality of services offered, their main distinction from critical defects is that the failure may be circumvented with some conservative policies. Finally, *minor* (severity 3) defects indicate features that would be nice to have to improve the service or to facilitate network operations. Problems that are classified as having severity levels 1 or 2 using IEEE 1044.1 correspond to the category of critical (or severity 1) problems using GR-929. The reason for this difference is that GR-929 originates from the practices of telephone operators who have to meet stringent regulations with less predictable or controllable events, which is not the case for the typical private (enterprise) networks.

For a historical perspective, Table 8.5 shows the Bell System classification of network defects [Rey, 1983, p. 748]. Evaluation of the impact of a defect requires the integrated views of many parties, project team, vendor organization, network operations, account executives, product managers, and representatives of the external customers, with the objective of converging on a common definition. Typically, this is achieved through a defect

Table 8.4 Description of Severity Levels for Defect Classification (Section A.16 in IEEE 1044.1-1995)

Level of Severity	Status of Service	Description
1	Urgent	The failure causes a system crash or unrecoverable data loss or jeopardizes personnel.
2	High	The failure causes impairment of critical system functions, and no workaround solution exists.
3	Medium	The failure causes impairment of critical system functions, though a workaround solution does exist.
4	Low	The failure causes inconvenience or annoyance.
5	None	None of the above, or the anomaly concerns an enhancement rather than a failure.

Table 8.5   Defect Seriousness Classification in the Bell System

Type	Class	Defect Description
Major	A	1. Will surely cause an operating failure of the unit in service 2. Will surely cause intermittent operating trouble 3. Will render unit totally unfit for service 4. Is apt to cause persona injury or property damage under normal condition of use
	B	1. Will probably cause an operating failure of the unit in service 2. Will surely cause trouble less serious than an operating failure, such as a substandard performance 3. Will surely involve increased maintenance or decreased life 4. Will cause a major increase in installation effort 5. Has extreme defects of appearance or finish
Minor	C	1. May possibly cause an operating failure of the unit in service 2. Is likely to cause trouble of a nature less serious than an operating failure, such as a substandard performance 3. Is likely to involve increased maintenance or decreased life 4. Has significant defects of appearance or finish
	D	1. Will not affect operation, maintenance, or life of the unit in service 2. Has minor defects of appearance, finish, or workmanship

review board—or modification request review board (MRRB). The advantage of a collective view is that events are not taken as isolated incidents. At the end of these discussions, the initial categorization of the defect can be confirmed or changed.

Appraisal

Appraisal of the quality involves evaluation of the system characteristics before service turn-up. Tests fall into five categories [Ali, 1995]: unit or module tests, functional tests, integration test, systems (or software) quality assurance (SQA) tests, and customer's acceptance testing. In incremental innovations, the operational profile of the equipment is well understood. This profile captures the various ways the network equipment will be used for the intended service and how they will interact with the rest of the systems used for operation and administration. Manufacturers have all the necessary information to conduct the first four test categories to verify that the system has stabilized and is fit for use. The network operator follows with a short acceptance test tailored to its environment, as illustrated in Figure 8.2.

In the case of new or emerging technologies, there are two complications. First, the operational profile is not known because it depends on the service features as opposed to the features of the product itself, the network engineering rules, and the network management scripts and tools, as well as methods and procedures implemented to run the network. In other words, the service features are often a subset of the total product capability. Second, the delivered product is not defect-free. Even when vendors implement the best current practices in software development (e.g., code inspection) and testing, the efficiency of defect removal is about 30% on the average, with a peak of about 65% [Jones, 1994, p. 445]. Given that, and if the operator still requires a reliable service, its role in quality assurance has to extend as shown in Figure 8.3. The implication is that the team involved in service

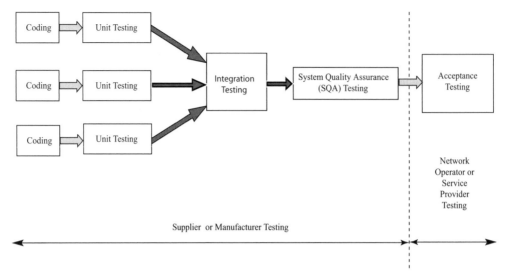

Figure 8.2   Software development and maintenance cycle for incremental service innovation.

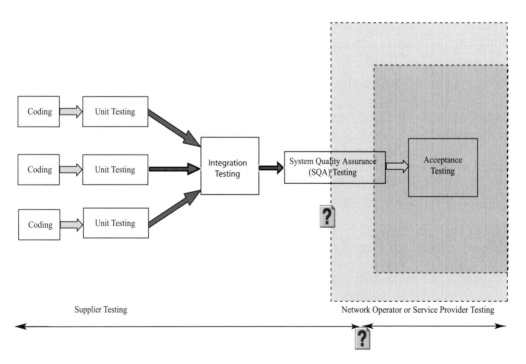

Figure 8.3   Role of telecommunications service providers in quality testing for nonincremental innovations.

development should include in its quality plan the task of verifying the reliability of the features that are needed for the service offer.

It should be noted also that start-up vendors, in particular, rush to establish market presence even at the expense of thorough testing; thus, for all practical purposes, they deflect the cost of quality assurance to their customers. Therefore, for nonincremental innovations, vendors determine the readiness of their products differently from network operators. Ideally, the network operator and the equipment manufacturer should define a joint test program whereby the network operator would provide the manufacturer with specific configurations or potential interactions that could take place in the live network such as cascaded failures, upgrade/downgrade scenarios, stress testing, and responses to extreme inputs. This gives the manufacturer the necessary information to make the changes as early as possible.

To avoid unnecessary expenditure in appraising the product, it is recommended that the project team examine the vendor's own data on the growth of reliability before embarking on the acceptance program. This is a run chart with the vertical axis representing the number of critical and major defects that the vendor has uncovered, and the horizontal axis represents time (calendar, man-hour, machine execution). If the defect discovery rate grows linearly with time, as shown in Figure 8.4, the project schedule should be reconsidered to reflect the need for continued testing. Additionally, the executives should be apprised to reevaluate the worth of the project, given the inevitable project overruns. A rigorous review will determine whether to cancel the project and reallocate the resources to more promising alternatives. If the product appears to have stabilized, the vendor data can also be used to estimate the extent of acceptance testing, provided that the analysis is restricted to those features that the operator is interested in and not all the capabilities of the product. Maximum likelihood estimates (MLE) can be obtained to determine the additional test duration for achieving a pre-specified level of defect coverage and a projected time for test termination. For example, the curves shown in Figure 8.5 suggest that defect discovery rate has tapered off. The statistical techniques presented in the Appendix suggest that about 150–200 additional problems may be uncovered in 300–400 days. With

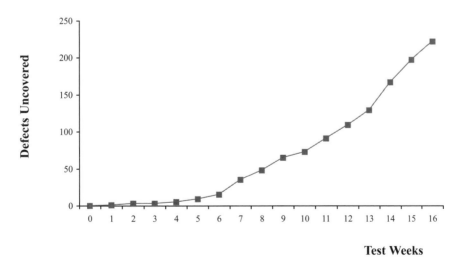

Figure 8.4 Vendor defect data that are still growing linearly with time.

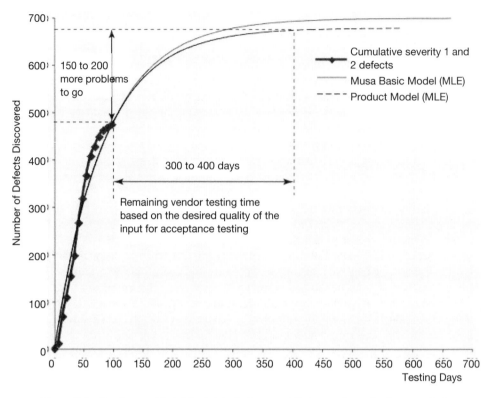

Figure 8.5   Vendor-provided defect data that suggest that acceptance testing can start.

such estimates, the testing plan (scope, resources, schedule) can be elaborated, including ways to compress the appraisal schedule. Without such a quantitative assessment, the sponsors may question the time and resources allocated to the quality control activities.

Schedule Compression

The length of the acceptance testing depends on the degree of collaboration between the network operator and the manufacturer. As explained earlier, the first approach consists in providing the vendor with specific guidance on what to test. This increases the coverage of the vendor tests and allows the vendor personnel the chances to observe and correct some of the problems before the product is frozen.

Another method is for the operator and vendor to carry a joint testing program. This cooperation should take place are the right time of the product cycle. In a premature collaboration—that is, before the product has started to stabilize—the churn could even delay the vendor's progress. If the joint testing is late in the development cycle, then the problems discovered will have to be postponed to further releases. The vendor should supply additional details on the testing results, including a list of known problems, the various changes that were made to previous releases, and detailed instructions on the product operation.

Each time a new release arrives, some tests are routinely conducted to ensure that the enhancements made into the previous release (to add new features or to correct faults) do not destroy or damage working functions. These well-honed and repetitive tasks that are carried from one release to the other are called *regression tests*. It is often advantageous for the operator to develop a library of automated regression tests to executed for each release, thereby shortening the testing time significantly or increasing the efficiency of test resources usage (for example, tests can be run autonomously at nights and weekends). Automated tests are also more reproducible. For automation to be successful, however, the test team must gain familiarity with the test environment (tools, equipment under test, etc.) before of improving efficiency through test automation

Evaluation of Testing Progress

Some of the parameters used to track progress are:

1. The number of test cases executed, passed, failed, and blocked, as shown in Figure 8.6. Blocked tests are those tests that could not be executed, either because of another problem or because of unavailable equipment. Failed tests guide the regression testing of the next maintenance release.
2. Progress per feature as shown in Figure 8.7. This could direct shifts in service priorities by delaying features that could not be terminated in a later phase.
3. Percentage and number of test cases that have passed.
4. Number of unresolved critical or major defects.
5. Number of defects without root cause analysis: These identify technical/operational problems that could not be solved with the available resource.
6. Turnaround time for defect resolution as a way of evaluating the vendor's responsiveness.

To predict the calendar date for completing the whole testing program based on the actual rate of progress, the earned value technique presented in Chapter 5 has proven to be very useful. Unacceptable schedule slippage would trigger a reexamination of the project's value.

When to Stop Testing?

Improvements in the quality of the software are estimated by observing the time series of failures during the system test. Technically, the observed rate of software failure rate in the field is different from the defect discovery rate in the testing environment. However, defect tracking during the system test gives a good indication of the risks associated with the software release, provided that either the testing profile is designed to correspond to the field operational profile or the test cases provide sufficient coverage of the expected field operations. Armed with these data, it is possible to define a stopping criterion that balances the cost of continued testing with that of deployment and uncovering defects during live service.

The test stopping rule is derived from the inequality (2). The decision depends on several factors such as the following:

- The vendor has fixed all critical defects discovered so far and all major defects that do not have acceptable workarounds.

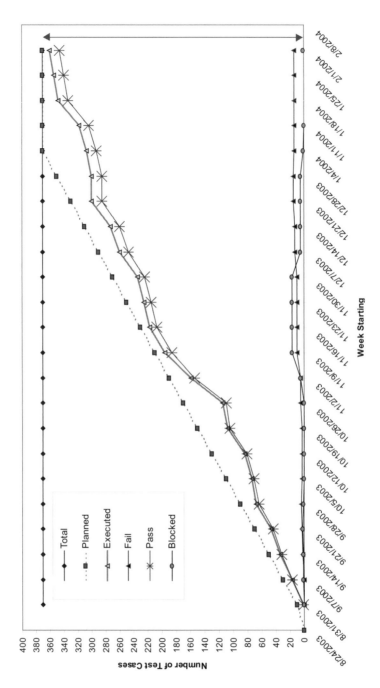

Figure 8.6 Tracking progress using test cases.

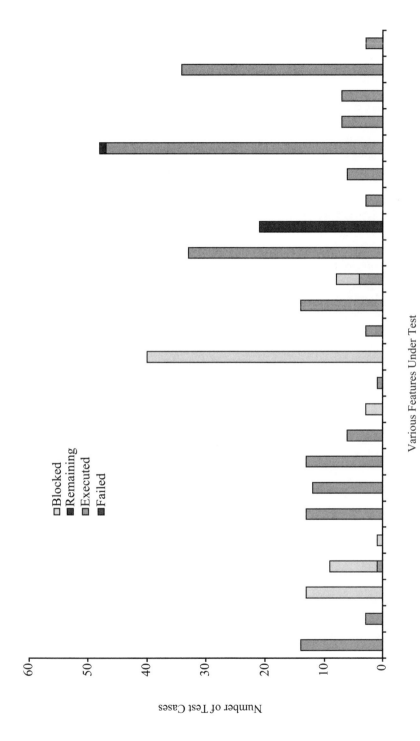

Figure 8.7 Tracking progress per feature.

- The rate of finding defects tends to zero and all tests have been covered. Specifically, the cost of continuing testing to discover the next problem (as estimated with the statistical techniques described in the Appendix) is higher than the cost of discovering that problem in the field.

- Alternatively, a less strict criterion would be that after all tests have been executed, the expected number of critical or major problems is less than a predefined number.

Figure 8.8 shows the evolution of the probability density functions and the cumulative probability as more testing are conducted. The way that they are calculated is explained in the Appendix. The cost of early deployment can be estimated as explained on page 137 (e.g., using the figures in Table 8.1) for data networks. In general, the probability density for a product that can be released should look like what is shown in Figure 8.9.

VENDOR MANAGEMENT DURING THE TESTING PROGRAM

Problems discovered during acceptance testing typically warrant "reworks" to bring the defective or nonstandard conforming element into compliance. In the general case, an existing version of the product may be already deployed in the field, while the next release is being tested for certification. Figure 8.10 depicts the information flow between the service/network provider and the manufacturer. It is seen that there are two independent paths for reporting problems to the supplier: from the acceptance test laboratory and from the field. Field reports represent complaints of internal and external customers. When the equipment is responsible for the field problem, the vendor has to have all the necessary information to duplicate it. Field failures drive improvements in the vendor's internal

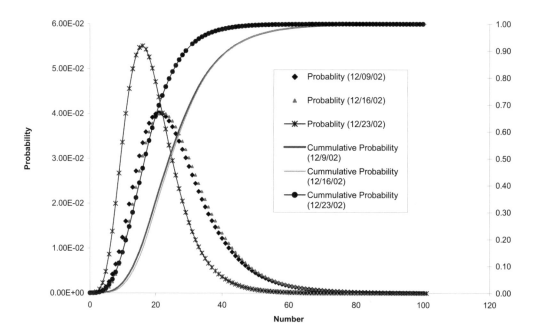

Figure 8.8 Evolution of the probability densities and distributions of a product.

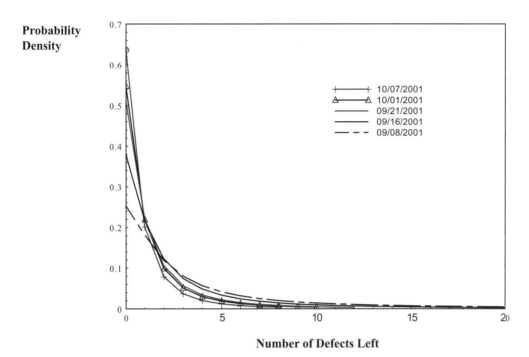

Figure 8.9 Probability density of the remaining defects before deployment.

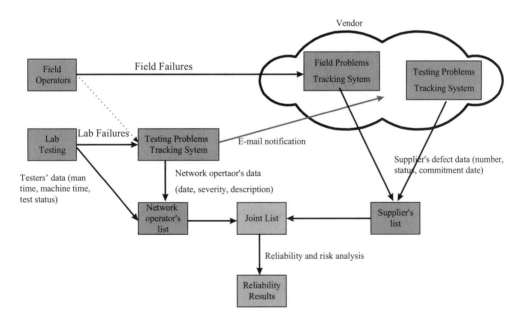

Figure 8.10 Information flow in the testing environment.

processes and the operator's quality assurance program: training of OSWF, acceptance testing, and so on.

Immediate notification of critical or major problems is needed; minor problems are usually postponed as per the agreement between the vendor and the customer. These data elements in the various databases are tracked and updated, for example, as the resolution is progressing or if there are changes in the defect classification. Urgent fixes arrive in an emergency release called "patch," which imposes an additional round of testing to verify that the rework did not provoke additional defects. Clearly, many data elements need to be exchanged and synchronized between the vendor and network operator to have a consistent view of the problem and its resolution.

Tracking of the vendor performance takes into account the rates of defect discovery and their resolution. Determination of the root causes of troubles relies on the vendor diligence because the service provider or network operator does not have access to the internal structure of the equipment (software architecture is unknown). Also, the supplier documentation may be incomplete or lacking, which leads to incomplete test plans. The average age of defects for each class is a measure of the vendor's ability to keep up with the defect arrivals, the percentage of fixes that are delivered correctly to assess the quality of corrective patches, and so on.

One possible way of tabulating the vendor's commitment to fix defects per feature per severity and per priority can be shown in Table 8.6.

SUMMARY

Quality management of service projects is different from that of product development. Defects that arise during testing or in the field are classified according to their severity and their priority. The severity is related to the impact on the network operations, while the priority relates to the phased deployment of the service. This two-dimensional classification takes into account the views of both the operator and the end-user, which are not necessarily identical. Statistical models for the trends of error discovery rates allow us to predict the quality of the service and make informed decision on deployment. This is done by tracking the number of defects found and the test time. Ideally, the starting point for tracking would be the vendor's own data from when they start their system test in preparation of the release. These data can be used also to estimate the amount of effort spent in the acceptance testing program. If the probability of finding an excessive number of defects is high, then the level of concern about releasing the product should be raised. If the

Table 8.6 Tabulation of Vendor's Commitment

| | Priority 1 | | | | Priority 2 | | | |
| | Severity 1 | | Severity 2 | | Severity 1 | | Severity 2 | |
Feature	Committed	Uncommitted	Committed	Uncommitted	Committed	Uncommitted	Committed	Uncommitted
Feature 1								
. . .	. . .	. . .	. . .	. . .	. . .	. . .	. . .	. . .
Feature n								

estimated time to complete the testing is large, the concern should be about the scope of the service or the amount of resources allocated for quality evaluation. As in all project activities, this tool provides objective inputs to assist—but not to replace—engineering judgment and business acumen. As a tool for quality management, the analysis yields another indicator to be used in conjunction with other data.

APPENDIX

During the testing, the defect count is recorded as a time series with the machine execution time as the independent variable. From a quality and risk management viewpoint, this time series can be used to make inferences about the stability of the features being tested, the estimated test duration to reach the desired level of quality, and when to stop testing based on a cost–benefit comparison.

The theory of software reliability is well known [Lyu, 1996; Musa et al., 1987; Musa, 1999] and is not repeated here. Common models to track the defect discovery over time are based on a nonhomogeneous Poisson process (NHPP). There is, however, a slight but important difference between the models of interest to equipment vendors and those that help network operators. Vendors are interested in the quality of the product development process, while operators use models that describe a particular product. In "process" models, the parameter of interest is the population mean, while in "product" models, it is the realization of that random variable as the actual number of defects. From a service availability point of view, the aim of the statistical analysis is to determine the properties of the product at hand and not the expected value of the reliability parameters, as would be the aim of a product manufacturer. While the corresponding maximum likelihood estimates (MLEs) of all models are typically close in value—provided that they have converged—their behavior differs with respect to estimating and interpreting the probability distributions.

Consider a series of m data points $t_1, t_2, \ldots t_m$, each representing a time point when a failure is discovered. Note that the inter-arrival time of the failures can be expressed by

$$t_i \quad \text{and} \quad t_{i+1} - t_i, \quad \text{for } i = 1, \ldots, m - 1$$

We use the following notation:

- N is the total number of failures due to defects in the product under test
- $E(N)$ is the expected value of N
- $n(t) = m$ is the amount of defects found by time t
- $\mu(t) = E(n(t))$ denotes the expected number of defects found by time t
- ϕ is the constant component of the failure rate
- δ denotes the failure rate component that is learned about during testing
- τ_j denotes the observed inter-failure time of the jth failure
- t_j denotes the observed failure time of the jth failure.

Poisson Model

The number of observed defects over time is said to follow a Poisson process if the inter-discovery time is distributed according to an exponential distribution with a mean of φ; that is, that the probability density of the inter-discovery time is given by

$$\varphi e^{-\varphi t}, \tag{1}$$

that is,

$$t_1 \sim \exp(\varphi) \quad \text{and} \quad t_i - t_{i-1} \sim \exp(\varphi) \qquad \text{for } i = 2, \ldots, m. \tag{2}$$

In this case, $\mu(t) = \varphi t$ and the defect discovery rate (also called also the hazard rate or the instantaneous failure intensify) $\lambda(t) = \partial\mu/\partial t = \varphi$ is constant. This means that the number of defects increases indefinitely with the test time.

The Basic Model

The basic model—also called the Musa Basic Model [Musa, 1997; Musa et al., 1987]—is described as

$$\mu(t) = E(N)\left[1 - e^{-\left\{\frac{\lambda_0}{E(N)}\right\}t}\right]. \tag{3}$$

The failure intensity $\lambda(t)$ is given by

$$\lambda_{\text{Basic}}(t) = [E(N) - \mu(t)]\varphi = E(N)\varphi e^{-\phi t} = \frac{d\mu(t)}{dt}. \tag{4}$$

The failure intensity decreases exponentially with time and is proportional to the expected value of the defects—that is, the mean of the defects across all software releases produced by the same manufacturer. The assumption is that the method for software development and testing does not change very much from release to release. Of course, this assumption may not be valid if there is fundamental change in the company structure due to merger, acquisition, or rapid turnover of key personnel.

The Jelinski–Moranda Model

The Jelinski–Moranda model [Jelinski and Moranda, 1972] is product-centric where a series of step functions represent the discrete nature of the error discovery. The failure intensity after $n(t) = m$ failures is given by

$$\lambda_{JM}(t) = [N - n(t)]\varphi = N\varphi e^{-\phi t}; \tag{5}$$

that is, the failure rate is proportional to the current fault content of the product and remains constant between failure detection. In this model, the discovery rate $\lambda(t)$ decreases as a staircase function each time a defect is discovered with a constant step.

Assume that $(N - j)$ defects remain at time t. The time to next discovery is the minimum of a finite set of exponential random variables, which is exponentially distributed as well. The expected time to the discovery of the first defect is $\frac{1}{N\varphi}$. The expected time to the discovery of the second defect is $\frac{1}{N\varphi} + \frac{1}{(N-1)\varphi}$, and so on. Therefore, the umber of defects $n(t) = m$ found at time t satisfies the following relation:

$$\frac{1}{\varphi} \sum_{j=0}^{m-1} \frac{1}{(N-j)} \leq t < \frac{1}{\varphi} \sum_{j=0}^{m} \frac{1}{(N-j)} \qquad \text{for } m = 0, 1, \ldots, N. \qquad (6)$$

This equation can be use to estimate the time remaining to discover the rest of the defects after the discovery of m problems.

This model can be reformulated to manage the risk during the introduction of a new technology within a live network. After discovering m failures we continue testing for a time until the cost of discovering the $(m + 1)$th failure exceeds the cost of discovering a failure in the field. To do so, we calculate the probability distribution of the remaining failures using Bayes statistics before making a decision. Furthermore, we collect data from the field and combine it with system test data to improve our confidence. Compared with the traditional sequential analysis, this method allows an earlier decision with better estimates for the service reliability [Sherif et al., 2002, 2003].

Deployability

The difference between the likelihood that the data arise from one of the assumed models and a Poisson model is a measure for "deployability" of the product. Suppose that the parameter estimation is conducted at time t_e, which does not necessarily correspond to a failure, then when m failures have been noticed at times, the likelihood function in this case is given by [Musa et al. 1987, p. 318]

$$L(\beta; t_1, \Lambda, t_m) = f(t_1, \Lambda, t_m)P[t_{m+1} > t_e | T_1 = t_1, \Lambda, T_m = t_m\} \qquad (7)$$

where β is the vector formed by the model parameter. Using the property of exponential distribution of the inter-arrival times, Eq. (4) can be rewritten as

$$L(\beta; t_1, \Lambda, t_m) = \left[\prod_{i=1}^{m} f(t_i | t_{i-1}) \right] P[t_{m+1} > t_e | T_1 = t_1, \Lambda, T_m = t_m\}, \qquad (7')$$

The log-likelihood function for the Poisson model is

$$\ln L_0(\beta; t_1, \ldots, t_m) = m \ln(\varphi) - \varphi t, \qquad (8)$$

where t is the current time (machine execution or calendar time).

In the basic model, the estimation is at the time of the last failure. The log likelihood of the Musa's basic (process) model is given by

$$\ln L_1(\beta; t_1, \Lambda, t_m) = \sum_{i=1}^{m} \ln \lambda(t_i) - \mu(t). \qquad (9)$$

Note that the log-likelihood function for the basic model is the same as that of the Poisson model if the failure intensity (discovery rate) is constant—that is, when $\lambda(t_i) = \varphi$, and $\mu(t) = \varphi t$.

Similarly, the log-likelihood function for the Jelinski–Moranda model is

$$\ln L_2(\beta; t_1, \ldots, t_m) = m \ln(\varphi) + \sum_{i=0}^{m-1} \ln[N - (i - 1)] - \sum_{i=1}^{m} \{\varphi[N - (i - 1)](t_i - t_{i-1})\} \qquad (10)$$

with $t_0 \equiv 0$.

The reliability of the system after the mth failure is given by $e^{-B[N-(m-1)](t-t_m)}$. Therefore, because all failures are equally likely to occur and are independent of each other, we have

$$L[t_1, (t_1 - t_2), \Lambda, (t_n - t_{n-1})] = \prod_{i=1}^{n} \varphi[N - (i-1)] \, e^{-\varphi[N-(i-1)](t_i-t_{i-1})} \tag{11}$$

and the log-likelihood of the modified Jelinski–Moranda model becomes

$$\ln L_3(\beta; t_1, \ldots, t_m) = m \ln(\varphi) + \sum_{i=0}^{m-1} \ln[N - (i-1)] - \sum_{i=1}^{m} \{\varphi[N - (i-1)](t_i - t_{i-1})\}$$

$$- \varphi[N - m](t - t_m) \tag{11'}$$

with $t_0 \equiv 0$.

The product is deployable when the test indicates that a model is a better fit than the Poisson model and the higher the ratio, the better. Let

$$\gamma_m = \frac{L_i(t_1, \ldots, t_m)}{L_0(t_1, \ldots, t_m)}, \qquad i = 1, 2, 3 \tag{12}$$

with $i = 1$ for the basic model and $= 2$ for Jelinski–Moranda model and 3 for the modified Jelinski–Moranda model. One possible test used to indicate that a model fits the observed data better than the Poisson model is

$$\ln \gamma_m = \ln \frac{L_i(t_1, \ldots, t_m)}{L_0(t_1, \ldots, t_m)}, \qquad i = 1, 2, 3 > 3 \tag{12'}$$

Thus if the deployability figure is 20 (3 in the natural logarithmic scale), the corresponding model is more likely than a Poisson model. Furthermore, the higher the deployability, the better the estimates of the number of the remaining problems to be found. The normal behavior that Figure 8.11 illustrates is that the defect find rate increases with time and then starts to abate, as the reliability of the product increases and most problems have been fixed.

When the defect find rate does not decrease—that is, it remains steady or even increases—the product reliability does not increase with additional testing and the product cannot be deployed. Consider the case of the failure data shown in Figure 8.12 and the corresponding deployability illustrated in Figure 8.13. Assuming that the testing is done correctly, the observed behavior provides evidence for a fundamental flaw in the software development process. Possible reasons could be deficiencies in change control or flaws in the software architecture (for example, the use of global variables that are overwritten unintentionally when fixes are made). Ultimately, if the reliability does not improve, one possible way out would be to restrict the features deployed to those that are known to work or, more radically, to reject the whole release.

Learning Effect with the Yamada Model

The above discussion assumes that the defect discovery rate starts in a linear fashion at the beginning of the acceptance test. In many applications, however, the discovery starts

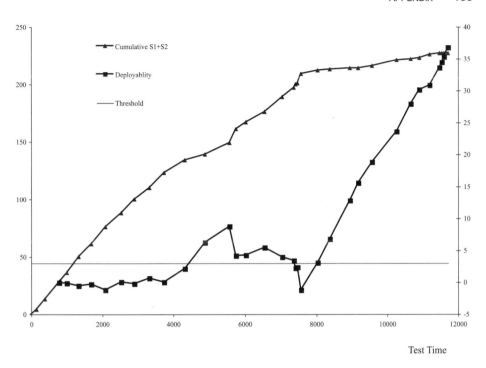

Figure 8.11   Evolution of a product deployability.

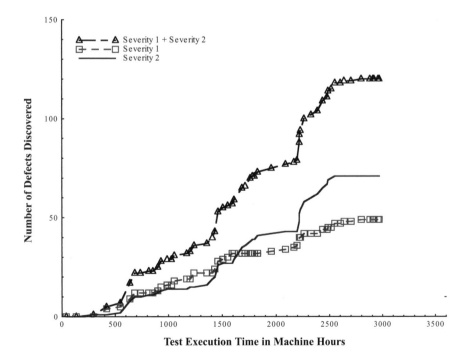

Figure 8.12   Observed failure rate for a product.

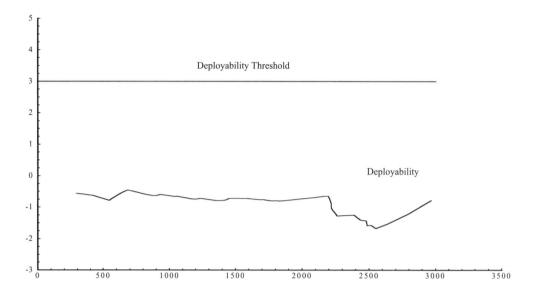

Test Execution Time in Machine Hours

Figure 8.13   Deployability of the product with the failure rates of Figure 8.12.

at a slow rate and then "speeds" up before it decreases steadily. Such a behavior produces an "S" tracking curve that reflects the "learning" by the test team. To account for "S"-shaped curves observed during such tracking, Yamada et al. [1983] introduced a nonhomogeneous Poisson process with a failure intensity that has the functional form of a gamma distribution with a shape factor of 2, which leads to the following equation:

$$\lambda(t) = [E(N) - \mu(t)]\frac{\delta^2 t}{1 + \delta t} = E(N)\delta^2 t e^{-\delta t}; \tag{13}$$

that is, the failure rate starts at zero and peaks after $(1/\delta)$ units of time.

The log-likelihood of the Yamada model is given by

$$\ln L_4(\beta; t_1, \ldots, t_m) = \ln \lambda_0 \sum_{i=1}^{m}(y_i - y_{i-1}) - \sum_{i=1}^{m} \ln(y_i - y_{i-1})! - \lambda_0[1 - (1 + bt_m)e^{-bt_m}]$$

$$+ \sum_{i=1}^{m} (y_i - y_{i-1}) \ln[1 - (1 + bt_{i-1})e^{-bt_{i-1}} - (1 + bt_i)e^{-bt_i}] \tag{14}$$

with $y_0 \equiv 0$ and $t_0 \equiv 0$.

This model has been adapted as well for service offers [Hoeflin and Sherif, 2005].

9

VENDOR MANAGEMENT

THE IMPORTANCE OF VENDOR MANAGEMENT

The success of most telecommunication services projects relies on a strong collaboration of vendors and buyers to create knowledge particularly early in a technology or a product life cycle. This was illustrated in Chapter 8, where it was shown that during the acceptance phase, the vendor and the network operator share their experiences to improve their respective offers. Large infrastructure projects, such as the establishment of undersea cables, provide a good illustration of the complexities in vendor management. In such projects, technical teams from several operators work together to come up with the specifications, devise a network architecture, and select the equipment suppliers. Legal teams establish the legal and commercial provisions for procurement and operation within the respective countries. Common logistics teams participate in the procurement decisions as well as in the installation and commissioning of the equipment. Financial teams establish jointly the pricing structure and develop accounting procedures and revenue sharing formulae. We see from the above that vendor management for telecommunication services goes beyond procurement—that is, beyond the set of processes designed to acquire goods and services from outside the performing organization [PMI, 2000]. This is elaborated in the next section.

VENDOR MANAGEMENT VERSUS PROCUREMENT MANAGEMENT

There are many similarities between procurement management and vendor management. Their scope is to obtain goods and services for a project in accordance with the technical requirements within the constraints of the schedule and expense limits. Both relate to the assessment, selection, administration, and evaluation of supplier relations. They aim at controlling the cost and quality of the transactions between a buyer and a seller. In each case, the first step is the definition of the requirements and the development of the statement of work and the product specifications. Furthermore, in both cases, evaluation of the supplier's performance depends on specific objectives regarding delivery intervals or order defect rates. However, vendor management and procurement management differs in some important aspects.

In general, procurement management focuses on the business decisions of making or buying as well as the associated contractual aspects. It is concerned with the efficiency of the supply chain in terms of tracking costs, auditing invoices, and integrating all pertinent data through e-business and web collaboration tools. In the case of government projects, there are laws and regulations that draw the boundaries of communications between the buyer and the supplier. This impersonal approach may be also suitable when the relationship is solely contract-based, short-term-oriented, and at arm's length, such as for products in the area of maintenance, repair, and operations (MRO). Here, good procurement practices can increase profitability by seeking quality suppliers and taking advantage of quantity discounts. Even for the procurement of strategic products in telecommunication services, such as network elements or network management systems, the approach may be useful when products are mature or standardized, such as in the case of incremental innovations.

Vendor management goes beyond the contractual or legal obligations, because it relates mostly to knowledge creation and sharing among the contractual parties during service development. This means that vendor management is more sensitive than procurement management to the context of the innovation, such as the phase of the technology life cycle and the type of innovation. In particular, when products are acquired early in their life cycle, many details regarding their capabilities and limitations are not known. Improvements result through several iterations with the collaboration of early adopters in areas that can be technical or commercial and extend across divisional or firm boundaries [Jin, 2001].

In short, vendor management goes beyond the contract management and covers the need to communicate across firm boundaries to help the creation of new knowledge. Thus, Deming's fourth point is very appropriate in vendor management: *"End the practice of awarding business on the basis of price tag. Instead, depend on meaningful measures of quality along with price. Move toward a single supplier for any one time, on a long-term relationship of loyalty and trust."* Vendor management is most appropriate for strategic goods directly related to the service creation and would be an overkill for indirect goods, such as those related to maintenance, repair, and operations (MRO). Table 9.1 summarizes the differences between vendor management and procurement management.

ACQUISITION PROCESS

ISO/IEC-12207 describes a generic acquisition process consisting of the following steps:

- *Need Definition and Procurement Planning.* This is the process of identifying which project needs will be procured from the project organization (i.e., outsourced). This involves a make-or-buy analysis and selection of the type of contracts.

Table 9.1 Differences Between Procurement Management and Vendor Management

Characteristics	Procurement Management	Vendor Management
Purpose	Minimize transaction cost	Maximize knowledge creation and retention
Nature	Impersonal and at arm's length	Relation-specific
Dependence on the technology life cycle	No	Yes
Areas of application	All types of innovation; covers all types of goods (strategic as well as indirect goods)	Supplemental for radical, architectural and platform innovation; for strategic goods only
Performance metrics	Delivery intervals and order defects rate	Delivery intervals and order defects rate
	Invoicing accuracy	Invoicing accuracy
		Innovation to improve the service offer
		Response to new feature requests
		Readiness to collaboration and information sharing

- *Requirements Definition.* This is the process to define the technical and business specifications of the product. The result is procurement documents used to solicit proposals from the prospective sellers. The principle here is to have a sufficient variety of vendors to be able to have a good selection. Typically, the term "bid" or "quotation" is used when the selection is price-driven while the term "proposal" is used when nonfinancial considerations are used (e.g., quality of the product, standards used, etc.). In particular, the following terms are used in the telecommunication services area: request for information (RFI), request for quotation (RFQ), and request for proposal (RFP). The evaluation criteria are often included in the procurement documents.

- *Preselection (Short Listing).* After the RFP has been issued, the bidders prepare their bid proposals. This is the stage where the evaluation criteria are used to rate or score the proposals and select a short list of vendors for additional scrutiny. This information is obtained through pre-bid vendor's presentations or site of the procedures, and so on.

- *Evaluation (Due-Diligence).* Bids are evaluated for ranking the vendors proposals. Assessment covers both technical and nontechnical factors such as technical, political, and market risks.

- *Supplier Selection.* Selection is usually based on the technical evaluation of the bids as well as other strategic and political considerations.

- *Contractual Agreement.* This involves the procurement and legal departments to define the contract terms that the selected bidder will have to comply with and agree to, the payment schedule, and so on.

- *Contract Management and Operation.* Contract administration is the process of ensuring that the seller meets its contractual obligations. It covers purchase orders, service orders, vendor contracts, and leasing agreements that would be needed to track the vendor performance or to reacquire equipment and services in the event of a disaster to restore the service configuration.

- *Closeout.* When all the requirements have been met, the project is closed. Final payment is contingent upon verification that all contracted work has been completed.

Evaluation of the Formal Solicitation Process

The formal solicitation process (e.g., RFPs) is typical for major contracts. However, in commercial projects, management may perceive that the formal solicitation process is wasteful and may occasionally request to skip it to reduce cost and to shorten the interval to market. The solicitation process may be sacrificed when the schedule needs to be compressed (*crashing*). This decision, if not well-managed, can create suspicion of favoritism with respect to the vendors that have not been selected and can give a wrong message on corporate integrity to the internal staff.

The telecommunications boom of the late 1990s saw tangled relationships that led to many excesses. One scheme explained in the *New York Times* would work as follows. A start-up manufacturer would offer executives of telecommunications operators some shares of its initial offerings as a "friend of the company." Next, that operator would announce a deal to buy some equipment from that start-up after it had gone public. Just the announcement of the deal would raise investors' expectations in the future of the start-up, and their acquisition frenzy will drive up the start-up stock prices, often multiplying it 10 or more times. The executive would then sell the shares before the stock collapsed when reality sunk in, thus pocketing the difference [Morgenson, 2002]. This example underlines why company guidelines should make the selection process transparent by clarifying when it is possible to skip the formal solicitation (in government contracts, this is specified by law). A company-wide policy is also needed to ensure the consistency of the evaluation across reviewers.

In general, the advantages of following a formal process are:

- Building a consensus within the firm concerning the service requirements and documenting the related decisions
- Learning about all potential vendors instead of focusing only on those that are known
- Increasing technical knowledge of the staff on the potential evolutions of technology, products, and markets.

Vendor Selection

Vendor selection is an important step because it determines the overall quality of the service offer in terms of reliability and features. Business criteria as well as technical criteria should be made explicit so that there is a transparent evaluation of the tradeoff between a short interval to market and the time to scale and to profitability.

There may be nontechnical reasons for vendor selection. However, short-term solutions to reduce the time to market without considering the future may not be scalable or efficient in the long term. The benefits and costs of such tactical solutions should be evaluated before taking a decision. Technical shortcomings will not go away and have to be considered before making the customer's commitment. Otherwise, the problems will be uncovered in the field, which may cause an embarrassment. Or, they may delay the introduction of the service to correct them. There should be a clear tradeoff analysis between an early introduction of a service with a product that will have to be replaced later versus getting the right quality from the first time. Recognition of this tradeoff may be hard when the firm's upper management is only considering the short term to satisfy the financial markets, because growth, scale, and reliability fit in a long-term perspective.

Top-down (i.e., strictly political) vendor selection is not conducive to long-term quality because it neglects technical evaluations and feedback from the field based on operational experience. In particular, in any telecommunications network of a reasonable size, the introduction of new equipment can cause incompatibilities or disrupt existing operations. Therefore, the technical evaluation teams should have the expertise to assess the technology and evaluate products, including the risks of incompatibilities and their effect on network reliability and operations.

Contract Type

Contracts are legal instruments by which an organization acquires products or services from outside sources. They define the roles and responsibilities of each party, the process of monitoring the contractor's performance, and the rules for rewards and penalties.

Without going into unnecessary details which can be found in many project management textbooks such as Kerzner (1998, p. 682], there are two basic forms of project contracts: fixed-price (lump sum) and cost-plus fixed feed (time and material). The first kind is suitable when the end result is known in advance, and the core project team has sufficient expertise on the item under contract. The specifications are detailed and eliminate any uncertainty concerning the technical characteristics, the cost, the schedule, and the quality. According to the second type of contracts, the payment to the contractor corresponds to the amount of time for labor and the equipment and material that were used plus a fixed fee. The contractor has now a large amount of freedom to experiment and change the scope of the activities while the project sponsor incurs a large amount of uncertainty regarding the cost.

Based on the data collected from 110 defense development projects, it was found that under low and moderate technological uncertainty levels, such as for incremental and architectural innovations, both the supplier and the customer are better off using fixed-price contracts. In contrast, cost-plus contracts are more beneficial when technological uncertainties are large, such as in the case of radical and platform innovations [Sadeh et al., 2000]. This finding is mapped onto the technology life cycle in Figure 9.1. Thus, in a staged development, cost-plus contracts are more suited to the feasibility study and the preliminary design while fixed-cost contracts are more beneficial in the phases of full-scale design and development.

VENDOR TYPES IN TELECOMMUNICATIONS SERVICES

Recall that telecommunication services are built on a networking infrastructure with its operations support systems and are operated with specific methods and procedures for engineering and operations. As a consequence, vendors can be of several kinds:

1. Technology vendors, for hardware, software, or both.
2. Connectivity vendors to fill in gaps in the end-to-end links of customers. Typically, these are bandwidth vendors that ensure connectivity to a specific region or country or that ensure trans-oceanic connections, and so on.

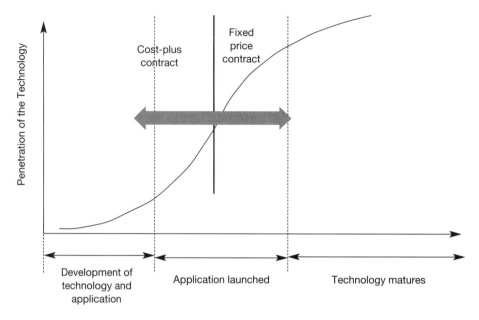

Figure 9.1   Contract type as a function of the technology life cycle.

3. Service vendors to support operations such as installers of equipment, billers of end-customers, training, on-site support for remote locations, and so on.
4. Consultants to improve the workflow and reduce costs.

The development of a typical telecommunications service relies on the simultaneous management of all these types of vendors. Consider, for example, the case of a disaster recovery service. In this service, the provider offers its customers the possibility to recover from massive failures by replicating the customer's data centers. To do so, however, it has to buy the necessary equipment and has to rent the necessary infrastructure from one or several network providers. Each network provider, in turn, designs a network configuration with preassigned (but inactive) backup ports and access circuits for each customer's data center. The end-customer designates which circuits that will be activated to ensure that mission critical applications are minimally affected by the failure. Activation of the disaster recovery plan is triggered when the customer reports to the service provider a site failure and requests reconfiguration. The disaster recovery service provider, in turn, calls the infrastructure network provider to effect the change.

In the following, we will treat all vendors together unless explicitly indicated.

VENDOR EVALUATION

A general approach to evaluate a vendor's technology on the basis of a technology audit model was developed by J. Garcia-Arreola (1996) [Garcia-Arreola, 1996; Khalil, 2000]. The evaluation looks into the following six major categories:

1. *Technology Environment.* This factor examines the environment of the company and whether it fosters teamwork, creativity, and flexibility. The environment factors include the quality of leadership, what strategies are adopted, organizational structure, technology culture, human resource management, quality controls on the design and manufacturing, and the way people are trained to master new technologies. The service provider's main drive is to ensure the timely delivery of features to the market (e.g., scale efficiency features, improved reliability to eliminate large outages, minimize service disruptions, exceed service availability commitments). This can only be achieved if the vendor is committed to service provider's vision for quality.

2. *Technology Evaluation.* This factor examines the company's level of mastering of existing state-of-the art technologies as well as emerging technologies. This is done across the value chain from upstream research and development to downstream activities of marketing.

3. *Market and Competitors.* A profound understanding of the environment in which the company competes and its relationships with suppliers, customers, distribution channels, and competitors. The evaluation includes technical and cost such as cost per port. This evaluation is more difficult for radical innovation than for architectural innovations (or systems innovations) and incremental innovations, because they are built upon well-practiced technological capabilities.

4. *Innovation Process.* In general, the ability to bring an innovation to the market in the shortest time is as important as the innovation itself. However, in telecommunication services, the short time to market at the expense of quality of service is counterproductive because of the cost of defects repair and management once the product is deployed in a network.

5. *Support Functions.* This factor evaluates the performance of the various functional areas in the support of the service (e.g., sales, distribution, intervals for service order, management of returns and replacement of defective cards, turnaround time for fixing problems in the field, etc.), staging process to verify and certify new hardware from major upgrade and swap activities to test all equipment before shipping. What happens when there is a problem? What happens when the problem cannot be reproduced? Support includes telephone support, dial-in support (for remote assistance), and on-site support. On-site support includes but is not limited to
   - Education/training
   - Test planning and review
   - Consultation
   - Trouble resolution in lab testing or field operation
   - Assistance in software and hardware upgrades
   - Analysis for systems statistics for recommendations regarding optimization and performance improvements

6. *Acquisition and Adaptation of New Technologies.* This factor evaluates the way the company evaluates new technologies with respect to making it or buying it, the decision process with respect to capital investment, selection of alliance partners, and so on.

Figure 9.2 summarizes the structure of the technology audit [Khalil, 2000].

We consider now additional factors that are specific to equipment vendors or connectivity vendors, respectively.

Additional Criteria for Equipment Vendors

These additional criteria include checks of their design guidelines (e.g., design for manufacturability, testability, reliability), quality control for manufacturing process, certification process for third-party vendors criteria for software release and methods for control of software quality, code delivery intervals, and sales and after-sales support. It covers relationship with suppliers, subcontractors, and partners. Metrics to evaluate the overall quality of telecommunications equipment vendors are defined in the TL 9000 requirements from the QuEST Forum (www.questforum.org) [Liebesman et al., 2001]. These metrics are based on documents from Telcordia Technologies (GR-929-CORE) and from the American Society for Quality (ANSI/ISO/ASQC Q9001-1994). They cover the following aspects:

- Product quality.
    - For hardware, this can be measured by infant mortality failures of cards, the circuit pack return rates, and comparisons between hardware return rates and predicted failures in time (FIT).
    - For software, it can be measured by software reliability estimates (as discussed in Chapter 8) and frequency of corrective patches.
- Process quality
    - Quality of the ordering process, change to orders (ship to address or characteristics and effect on delivery dates, partial shipment, documentation, etc.).
    - Time to respond to trouble, time to escalation, time to provide root causes of failures, and turnaround time for defect resolution as measured by the average lifetime of an open modification request.
    - Percentage of releases or patches that are delivered on time.
- Vendor support:
    - Quality of help in the installation and configuration of equipment
    - Quality of training
    - Procedure to handle engineering change requests and engineering changes
    - Quality of documentation—in particular, completeness of documents that capture changes between successive software releases or hardware

Additional Criteria for Connectivity Vendors

Among the factors of interest in this case are the following:

- Geographic coverage
- Service lead time such as the interval to respond to quote requests, provisioning time, and time interval to modify a circuit
- Technical parameters such as availability and mean time to repair (MMTR)
- Ability to provide adequate support

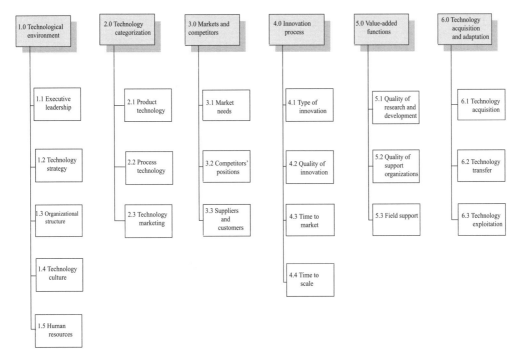

Figure 9.2 Technology audit structure.

COMMUNICATIONS WITH TECHNOLOGY VENDORS

During the development and the deployment of a service, there are many interactions among different members of either the vendor and the supplier. The information flow pertains to contracts, markets, and technical data regarding the equipment or real-life operating conditions.

Statement of Work

The statement of work should be clear and complete to avoid future problems due to ambiguities and to varying interpretations. The description should avoid jargon and unusual terms—unless clearly defined—to avoid misunderstanding and ambiguities. It should contain the following

1. Main deliverable
2. Other deliverables in addition to the product (designs, drawings, functional specifications, detailed software release notes, test results, etc.)
3. Tracking of the mortality rates of cards upon arrival and their reliability
4. The schedule under which the services or deliverables will be provided, including periodic reviews
5. The acceptance criteria for the product from the vendor

6. Division of work and the respective responsibilities of the supplier and of the buyer

7. Formal procedures for requesting change: who will make the request, how and when they will be made, who would approve them, and so on

8. In case of collaboration, how the knowledge generated will be treated

9. What is expected from the vendor in terms of support (as distinct from ongoing maintenance and support)

10. Location where the work is to be performed

11. The compensation in terms of fees and schedule

12. Delay penalties

Vendor Tracking

The network provider and the vendor collaborate in the development of the service. This includes joint scheduling of delivery dates and joint reviews at the technical, engineering, and marketing levels. Periodically, appropriate subject matter representatives from both sides would meet to assess the evolution of the market and the strategic response of each party, taking into account testing activities and problems reported from the field. In particular, in release planning meetings, both parties identify, review, and agree on a program plan for the next 6–18 months. For example, the supplier provides the operator with the functional specifications of the new features. In turn, the network provider may make operational data available to help the manufacturer conduct more realistic testing, such as failure scenarios to test the hardware and software to improve automatic fault detection and diagnostics. Finally, exchanges among the teams involved in network operations on the service side and customer support on the supplier's side relate to problems in the field. This channel of communication should include provisions to return and replace failed cards from central offices, a process often called return management authorization (RMA).

Tracking of vendor performance during service development may involve several metrics:

1. Accuracy and completeness of the documentation regarding the product architecture, test plans, test results, known problems, and so on

2. Use of best current practice in the product development process

3. Hardware reliability including mortality rate of cards on arrival

4. Defect resolution process: speed of response, failure rate of fixes, and so on

5. Support during deployment in the field

Figure 9.3 depicts the functional interfaces between a service provider and a technology supplier.

Partnerships and Virtual Organizations

Recent developments in technology and user needs have evolved telecommunication services offers into three service "spheres": the home, the enterprise, and on the road [Le Maistre, 2004]. The role of the service provider has become to deliver the content to the subscriber irrespective of the underlying transport and network technology—for example, by removing the boundaries between fixed and mobile services. Carriers cannot do this by

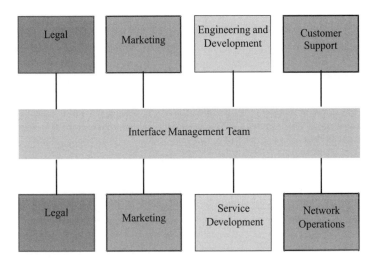

Technology Supplier

Network Operator

Figure 9.3 Possible communication channels between network operators and technology vendors for emerging services.

themselves and depend on partnering with many suppliers to create new services using multiple platforms and to deliver the content to any customer irrespective of the type of terminal they have. Partnering is a way to collaborate with suppliers, share experience, and learn together for mutual benefit in areas of engineering, marketing, subscriber service, and so on. We can see, then, that both parties are combined in a temporary team or a "virtual organization" tied by formal and informal networks. This does not mean that each company is "stripped to its bare bore competencies" [Gunn, 1994], but that they are forging a temporary alliance during the project. Furthermore, thanks to new information technologies, the virtual organization can be distributed geographically.

Management of such an organization demands a disciplined process to coordinate plans and to share the data needed for controlling the service development processes. Unmanaged channels will lead to confusion and misunderstanding if contradictory information and/or spurious cross-talk interfere with the channels, such as differing views on the requirements or development plans. Also, these exchanges are regulated by laws as well as the terms of the contract with respect to the protection of proprietary or third-party information (information about other vendors or other network operators that each party may know). Because the communication remains supervised by the management of both companies, the virtual organization remains an extension of existing hierarchical and functional networks.

It should be clear, then, that this type of organization is substantially different from typical supply management arrangements where the purpose is to reduce inventory and enhance response to market conditions through mass customization and just-in-time manufacturing and where the hierarchy of the existing organization is bypassed. The differences between virtual organizations in manufacturing and telecommunications services are summarized in Table 9.2.

Figure 9.4 shows the information exchanges in this virtual organization. The manufacturer coordinates the planning of its releases with the feature planning from the service provider based on market information or customers' inquiries. The exchanges cover feature definitions, test plans, operating procedures, and results—particularly statistical analysis of failure data and trends, as well as feedback from field data. To improve management of these interfaces, several telecommunication providers such as France Telecom and BT have reorganized their structure by putting their services and partnerships, rather than network infrastructure and internal research and development at the center of their decision process [Le Maistre, 2004]. It is too early to tell whether this structure is successful.

Metrics for Vendor Tracking During Acceptance Testing. Some of data to assess vendor performance in improving the reliability of the software release are

- On-time delivery of releases
- Number of problem reported
- Number of problems fixed and unresolved
- Frequency of corrective patches
- Percentage of fixes that are correctly delivered
- Turnaround time for defect resolution for problems discovered during testing in terms of
    - The average duration between the fix request date and the delivery commitment dates
    - The maximum amount of time that a modification request remains without a commitment from the supplier to fix
- Availability of root causes for problems discovered
- Availability of documentation regarding the changes between successive releases

Vendor's Handoff. Release notes from the supplier to the operator include information on

Table 9.2   Differences between Virtual Organizations in Manufacturing and in Telecommunication Services

Feature	Manufacturing	Telecommunication Services
Purpose	Lower costs by reduced inventory and faster response to orders	Knowledge generation
Organization	Nonhierarchical	Hierarchical and functional
Inputs	Customer order	Field problems or new feature requests
Existing hierarchies	Bypassed	Maintained

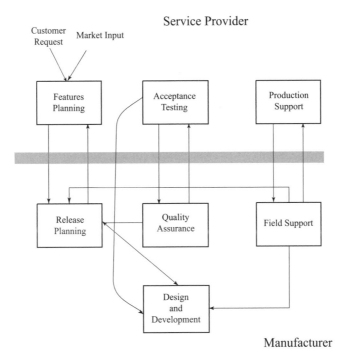

Figure 9.4 Information flow between telecommunication service providers and equipment vendors during development of emerging services.

- Release dependencies
- Interoperability between releases
- Differences between successive software releases, particularly changes that may affect the service
- Known problems
- Software reliability data

In a release commitment letter (RCL) the network operator describes what features have been accepted and which ones are lacking, an expected date for delivery of the missing items, and the agreed-upon method for tracking and resolving problems that may arise during operation.

Metrics for Vendor Tracking for Problems in the Field. The data are similar to the performance during acceptance testing but with an emphasis on urgency of response (since the problems affect customers traffic and potentially revenues):

- Number of problems reported
- Time to resolution, time to escalation, time to closure for field problems
- Time to provide root causes of failures and field defects
- Turnaround time for defect resolution

- Percentage of correct fixes that are correctly delivered
- Frequency of corrective patches
- Circuit pack return rates

RISKS IN THE MANAGEMENT OF TECHNOLOGY VENDORS

The contractual ties between the vendor and supplier may be sufficient for incremental innovations or spot contracting on the open market. Other types of innovations, for example, call for more collaboration in the form of alliances, joint product, and process development. We consider the following risk factors:

1. The technology life cycle
2. Vendor type
3. Supply disruption
4. Congruence of the supplier and the service provider plans
5. Standardization
6. Intellectual property and knowledge management
7. Inadequate field support

The Technology Life Cycle

In the case of mature technologies, stringent technical and contractual specifications with objective performance measures are usually incorporated in the contract. Penalty clauses may be requested in case the reliability/performance requirements are not met.

The risks increase considerably when the product is based on a new technology because of the many uncertainties involved and because the technology may be several years away from reaching its optimum capabilities. There are two questions here. The first is, How much more development is needed for the product to be ready? The second is, Can a service based on the current product be deployed to gain some experience with the technology?

As explained in Chapter 8, the review of the status during testing may indicate that the supplier is not capable of delivering the product within the required time frame. In the case of new technologies, a joint test program should be discussed. Risk mitigation strategies include working with the manufacturer to improve design and manufacturing process, develop and implement design guidelines, and put more stringent hardware quality control and audits on all cards. Common test plans will be developed, reviewed and agreed upon by both teams to focus on the issues that are blocking progress. The joint testing program will include a joint modification request review board and defined measures to track improvements for each release. It is important in this case to have insight into the quality of the software reliability by adopting special processing techniques.

Vendor Type

The risks in vendor selection vary depending on their track record (established or start-up). Technological discontinuities are creating opportunities for start-up companies that are risk-oriented and can take quick decisions. Unfortunately, they typically reduce the development cost and shorten the time to market by paying less attention to best practices in software development and quality testing. Furthermore, feature documentation may be inadequate or lacking, which leads to incomplete test plans. Defects may be hidden deeply within the software architecture. For example, the software can be monolithic with application level, operations systems level, and device driver code intertwined, even though modularity is necessary for reliability, maintainability, and so on. In other words, they pass on the cost of defect prevention and/or detection to the early adopters and subsequently, to the established companies that may acquire them. This explains why start-up vendors do not implement the best current practices such as use of reviews or code inspection. Note that even if these practices are implemented, the average efficiency for defect removal is about 30% and the peak is about 65% [Jones, 1994, p. 445].

Table 9.3 summarizes the main differences of start-up suppliers and large incumbent firms.

Risk of Supply Disruption

Because of the instability of the telecommunications market, there is a risk that a supplier may discontinue the product or that the supplier itself might disappear through mergers, acquisitions, or bankruptcy.

Even if the supplier survives, personnel reduction and successive reorganizations may cause delays or loss of expertise that can affect the product quality in terms of the feature set, the reliability, the implementation of best current practices in software development, and so on.

Congruence of the Plans for the Vendor and the Service Provider

The objectives of the network operator may be different from those of the equipment vendor.

1. The aim of the service/network provider may be just experimentation with the new technology to assess the interest of potential customers and to think about potential applications and prepare for the evolution of its service portfolio. The vendor's objectives, however, usually go beyond experimentation. The objective may be gathering technical information to improve the product or collect market data to expand market share or to monopolize the whole market. Thus, the service provider is merely one of a list of potential customers.
2. The goal of start-ups is to generate sufficient good publicity to whet the appetite of

Table 9.3 Comparison of Start-Ups and Large Established Firms in the Telecommunications Sector

Characteristics	Start-up	Large Established Firm
Flexibility and risk orientation	High	Low
Best Practices Implementation	Usually low	Usually high
Time horizon	Short-term	Long-term
Decision making	Fast	Slow and hierarchical
Quality of Documentation	Usually low	Usually high

investors before an initial public offering (IPO) of the company. The goal of the service provider is to offer a real service that is able to scale or to operate under extreme conditions with the same amount of quality. The vendor's organization may be in a state of flux after being acquired. No firm commitment will be made on bug fixes or on delivering new features. This may not be important for the vendor, particularly if it knows that they may not be able to survive without an infusion of cash. So there is a risk of supply interruption.

Lack of Standards

The presence of equipment from different vendors or different generations from the same vendor is inevitable in modern networks. In a complex network that is evolving with technology and customer needs, there is the danger of incompatibilities. The risks should be evaluated and a mitigation strategy defined.

In general, the mitigation strategy relies on two basic approaches: (a) have standardized interfaces and (b) work out the interoperability issues before deployment. In the latter case, a vendor test environment may be needed to duplicate the conditions in the operator's production network so that problems can be discovered at the manufacturer's laboratory before shipping the product.

Moving to alternate vendors is easier if standardized interfaces have been established. However, switching to a different supplier may be unavoidable if trend analysis indicates that the supplier has difficulty meeting its commitments.

Intellectual Property and Knowledge Management

Collaboration with suppliers and vendors and customers to develop new products or services involves significant exchange of ideas. Because of the intensive collaboration, there is the risk that proprietary information and trade secrets be divulged during operation. The value of intellectual property is to allow market positioning. During this collaboration, it is important to ensure that the network operator takes all necessary steps to protect all the parties involved. In particular, since a network operator often deals with several equipment manufacturers simultaneously, the proprietary information from one vendor must be blocked from going to another competitor. Also, the network operator should not give a supplier or a customer any reason to believe that it has waived its rights in its own intellectual properties. This is particularly important for information communications through letters, e-mails, or casual conversations.

Typically, during contract negotiation, the legal team inserts language with respect the obligation of each party to keep the other parties' information in confidence. The team may also insert some language corresponding to the rights of the developed work or to any potential patents that might ensure. However, these considerations cover essentially what can be described as codified knowledge. Also, in government or military contracts, it is possible to retain control of the technology by insisting that it not be sold to other parties without the funder's permission. Such an approach is typically too expensive in commercial environments.

There is another type of knowledge, namely, tacit knowledge or know-how that is embedded in the firm organization as a whole, or in a few key experts that the above legal considerations cannot protect. Tacit knowledge is based on the cumulative experience that has been developed over the years and that is exchanged implicitly or explicitly through direct contacts at lower levels. Consider the case of developing a network monitoring tools. The development cannot proceed without the network operator giving some detailed information on the way they detect problems and how they respond to them efficiently. This requires that the network operator reveal the variables to monitor, the thresholds for alarms, and the algorithms to treat all these variables. The information is then coded in a product to be used in the network. In general, the manufacturer can take that product or a derived product and sell on the market to other network operators that may be competing with the operator that has contributed to the development. Consider the case of an equipment evaluation where even though the equipment cannot pass the screening test, the evaluation itself is an important learning experience for the manufacturer. In all these situations, there is information transferred from one party to the other.

Inadequate Field Support

During the acceptance testing, the supplier's team may be unable to respond to the problems reported because of a different test environment (not the same amount of traffic, different configuration of the network element or the test equipment, etc.). When a problem arises during operations, the network operator would like response from the vendor as soon as possible. However, the vendor may not be capable of offering a 7×24 or may not have the resources to reproduce the reported problems. Vendor support is not the same in every country and may be unavailable in some regions, which is a problem for global or international networks. Having vendor's personnel on site in the testing lab or in the Network Operations Center can speed up issue resolutions, but this mitigation strategy is not always available.

Risk Mitigation in the Management of Technology Vendor

Managing vendor's risks depend on the size of the vendor and whether the vendor has organizational capabilities or masters the technology. The technology audit presented earlier in this chapter may be useful in evaluating the capabilities of the vendor. It provides a way to evaluate the status of technological assets of an organization in terms of its strengths, weaknesses, opportunities, and threats. It relates to the entire set of value-added functions in the firm, including product technology, production technology, service technology, and marketing technology.

The following additional steps may help alleviate the risks

1. Include properly baselined requirements of the service, even if some details need to be periodically updated through a change control process as the technology and the product mature. The requirements may be enhanced using operational scenarios to show interoperability situations with existing or future equipment. These cases help the vendor focus on the potential sources of problems when they do their own testing.

2. Include in the contract the best practices that the vendor should abide by in the development process. For equipment vendors, add the contractual obligations with major vendors to share tests plans and test results concerning the reliability of the product (hardware and software), Vendors will need to provide this information upfront as well as during the development of the product until generally deployed.

3. Hold the vendors to their contractual obligations in terms of quality, delays in delivery, and so on.

4. Organize periodic meetings at the management level and at the working level with

vendor teams to work jointly on reviewing problems and their resolutions, schedule freeze dates for code changes, code delivery dates, maintenance releases, upgrades, and so on.

5. Drive to standardize the interfaces in standard bodies. This allows the exchange of information among vendors in an open forum guided by anti-trust regulations that will enhance the overall quality of the industry.

CONNECTIVITY VENDORS

Types of Agreements Among Network Operators

There are several types of agreements among network operators depending on the services offered:

1. Interconnection agreement for the delivery of users traffic to areas or countries where the network operator has no access
2. Telehousing agreement to provide floor space for the network elements (switches, routers, etc.) and the associated spare parts in addition to other support functions such as:
    - Special physical preparations such as raised floors, air-conditioning equipment, DC power supply for the central office equipment, uninterrupted power supply (UPS), back-up diesel generator, access security, fire protection, and so on
    - Maintenance and routine testing of the site physical equipment mentioned above, including the keeping of proper maintenance logs
    - Fault reporting
3. Full service support through the following functions:
    - Participation in the pre-sales, ordering, provisioning of the product infrastructure and customer ports and access
    - Deployment of the network elements and network element management systems
    - Ongoing maintenance of the network equipment
    - Capacity planning including sparring, planning for growth, and deployment of line cards
    - Interface with equipment suppliers including the return management authorization (RMA) process
4. Forced agreement: The notion of an *unwilling supplier* arose from the deregulation of the telecommunications industry, where a company is forced by law to do business with another [Ward, 1998, p. 145]. For example, incumbent wireline carriers are obliged to enter into agreements with their competitor to offer them access.

Risks Management for Interconnectivity Vendors

In the case of interconnection agreement, commitments to customers are being made based on the commitments of other parties. Thus, the capability of a provider to meet its commitments to its customers in terms of delay and quality of service depends on the quality of its interfaces with other providers in addition to its internal operation. The risks of delay augment when the agreement has been forced by law such as when local access providers are obliged to offer access to their competitors.

Availability of services and their reliability as well as the type of service level agreement (SLA) depend on the country—for example, number portability or subscriber identification and numbering (i.e., telephone number). Legal systems differ in what type of encryption is allowed and what constitutes criminal activities. Rights of ownership of the content, right to privacy, responsibility of the carrier with respect to the content, different views of what constitutes security breaches and intrusions, and so on.

Service lead time depends on the bandwidth requested, the country, and the location within the country. This delay should be considering the making the project schedule.

SUMMARY

Textbooks on project management typically reserve a chapter on contracts and recommend that the project team seek support from specialists in the disciplines of contracting and procurement. Vendor management is needed in telecommunications equipment to cover the interactions that take place throughout the project life. In addition to management of the acquisition process and vendor selection, it covers management of the exchanges with the vendors in a virtual company environment. For emerging technologies in particular, the vendor and the network operator are bound temporarily to share skills, costs, and partial access to one another capabilities. There are many risks in the case of vendor management, and these have to be taken into account and managed. In the case of nonincremental innovations, collaboration of suppliers and network providers is essential to reduce the development time by sharing experiences and knowledge. Some of the pitfalls in vendor management are as follows:

1. There could be multiple interfaces between the vendor and the supplier with no centralized management so that issues will not be tracked.
2. Expectations on both sides beyond the contractual obligations are not clear.
3. Reasons of vendor selection that are not strictly technical may not be clear to the project team, thereby affecting the project moral. In particular, market-driven selections may force tactical solutions that are not conducive to long-term strategic growth.
4. The status of the knowledge gained through the exchange of information is ambiguous.

10

RISK MANAGEMENT

A project risk is an event that disrupts the execution of a project and reduces its chances of meeting its objectives of quality, cost, or schedule. A risk event is characterized by its root cause, the probability of its occurrence, and its financial consequences in the form of a loss function (e.g., the annual loss in business as a consequence of these risks; an estimate of the expected value of the loss; or a net present value of that loss). Using the cost as a measure helps bridge the gap between the technical and the financial sides of the project.

The contribution of risk management to the overall management of the project is that it provides a consistent methodology, along with tools to anticipate and quantify risks and communicate them to the various project stakeholders. Risk management has also a positive aspect on the overall process for project management because it encourages systematic collection of data across the entire organization to apply quantitative models to analyze project outcome.

KPMG, the professional services firm, has identified several categories of risk: strategic, operational, financial, regulatory, technological, and reputational [Orr, 2004]. As an integral part of the project plan, risk management consists of the following steps:

1. Identification of various root causes of risks
2. Risk evaluation to prioritize the treatment of various risks
3. Risk mitigation
4. Financing the residual risks

Risk management strategies must be tailored to each industrial sector as well as to the needs of individual companies. The purpose of this chapter is to present key ideas on how to manage risks in projects for telecommunication services. We focus on risks of techni-

Managing Projects in Telecommunication Services. By Mostafa Hashem Sherif
Copyright © 2006 The Institute of Electrical and Electronics Engineers, Inc.

cal origins or that are under the control of the project team. Even though telecommunication deals have shown some unsavory aspects bordering on fraud and corruption, the legal risks, while important for final success, are outside the scope of this discussion. Also, we do not consider extreme risks such as abduction of personnel, terrorism, problems due to fraud, and so on. A more detailed discussion of all these aspects can be found elsewhere—for example, Sirén [2003].

RISK IDENTIFICATION

Risk categorization and evaluation depend on many factors such as the type of innovation, the nature of organization, and the personality of the project manger.

Two common techniques to identify risks are *experience-based* and *brain-storming-based* risk assessment [Royer, 2000].

Experience-based risk identification involves expert assessment of the resources needed to achieve the project goals and the consequence of the absence or malfunction of any of these resources. The expertise can be personal—that is, that of the project manager, organizational, or sector-wide by benchmarking similar projects within the same industrial sector. Even if the organization itself does not systematically collect data on past project performance or does not encourage closure reviews, project managers should take it upon themselves to document their risk management experiences during the projects and proactively share them with other project managers. Obviously, this method is more appropriate for the case of sustaining innovations because of accumulated knowledge through successive projects.

Brainstorming-based risk identification is very useful when statistics on exposure to risk are scarce or when the innovation is disruptive so that benchmarking data are unavailable. In this case, the service architecture and the logical structure of the hazard model become the starting points for a collective exercise on risk evaluation and mitigation. For example, even though wireline telephony and wireless telephony provide end-to-end connectivity for voice, the risk model in each application is not the same because of the changes in the respective value chain. Customers in wireline telephony are residents in households, while the customer set of wireless telephony consist of individual users, mobile or sedentary. Furthermore, by changing the charging and billing models that were traditionally used in fixed-line telephony, wireless telephony introduced changes to the OSS and the workflow procedures for operations, administration, and maintenance.

Some organizations may have much experience introducing new technology and therefore understand how to deal with technology risks. Risk aversion depends on the type of risks, the cultural environment, and the personal traits of the project manager.

RISK EVALUATION

Risk evaluation requires a thorough analysis of (a) current and anticipated conditions, (b) the probability that these conditions arise, and (c) their likely consequences. The purpose is to help prioritize the risks to prepare better risk mitigation strategy by answering some questions such as:

1. How likely will the project be completed on time?
2. If likely to be late, how many days will the delay be? What are the contingency plans?
3. When is the project's greatest vulnerability to risks?

The results of the risk assessment can be used to tailor the way risks are treated using a risk map. The simplest risk map is a two-dimensional chart depicting the probability of event occurrence on the ordinate axis and the value of the corresponding loss function on the abscissa, as shown in Figure 10.1 [Barthélemy, 2000; Jablonowski, 2000]. The assessment of the probabilities and the loss depends on three factors: (1) the nature of the innovation, (2) the cultural environment in which the team operates, and (3) the personality of the project manager. The effects of these factors will be described later in this chapter.

Table 10.1 summarizes the risks classification using the above risk map.

The exposure to risk is the expected value of the loss, i.e., it is the probability of an event multiplied by the corresponding loss.

The worst case would be an event of very high probability and a high loss, as shown in the upper right-hand corner of Figure 10.1. Events with low loss and low probability can be documented as assumptions that can be tracked during the life of the project using metrics established for them. They can be tracked as assumptions [Royer, 2000]. All assumptions should be documented and tracked to ascertain their remaining validity using a monitoring metrics and agreed-upon thresholds to trigger a reevaluation of the plans or the use of a contingency plan. For example, the growth of reliability of a software release can be tracked in the system test program so that if it does not increase within 6–9 months for the start of testing, a jeopardy should be declared.

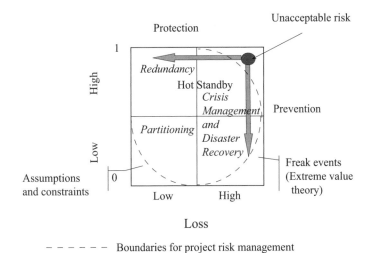

Figure 10.1 A simple risk map.

Table 10.1   Risk Classification

Risk Category	Description	Action
Unacceptable	High frequency and high loss	Bring the exposure to a man through a combination of prevention and protection including avoidance
Frequent	High probability of occurrence and low loss	Reduce the occurrence through preventive steps
Manageable	Average probability of occurrence and average loss	Protection through crisis management, redundancy, partitioning, etc.
Extreme	Low probability of occurrence and high loss (freak event)	Disaster recovery, also use Extreme value theory
Negligible	Low probability of occurrence and low loss	Consider it as part of assumptions and constraints

RISK MITIGATION

For each risk event, the mitigation strategy could be to reduce either the probability of the event (risk avoidance or prevention) or the corresponding loss function (risk reduction or protection), or a combination of both, as well as risk deflection.

Risk Avoidance

Prevention is the set of actions taken to reduce the probability of occurrence, including total avoidance by changing the project plan so that the risk event does not arise. Prevention should include an escalation procedure to engage the next level of responsibility if a road block or an obstacle has been identified through a monitoring process. The probability of occurrence of that event is reduced by either modifying the scope of the project (cost, schedule, requirements) or changing the way the implementation proceeds. In the limiting case, the event can be completely eliminated.

For example, the service architecture or the workflow procedures can be changed to prevent or reduce the chances that a given problem arises. Another example is that of tracking the vendor's progress using some objective metrics, such as the time it takes to correct problems discovered during testing, the estimated number of remaining problems in the software, and so on. If these measures fall below an agreed-upon threshold, then intervention may be warranted.

Risk Reduction

Protection consists of corrective actions to limit the consequences of the problem, such as following an emergency and in cases of a large-scale failure; these corrective actions include service restoration and protection.

Architectural modifications include the partitioning of resources in such a way that they are not all affected by the problem and are providing some redundancy. For example, sharing the same work package between two people protects against the absence of any of them. Similarly, the work can be partitioned among several geographic locations for security reasons in addition to continuity. In the same vein, continuity of network services can be enhanced by partitioning the network into two parts, each carrying part of the traffic on the assumption that sections that are geographically dispersed will not be affected simultaneously.

Protection by redundancy depends on the duplication of resources so that the second is used when the first fails. For example, to protect against loss of power supplies, uninterrupted power supplies can be used. In fault-tolerance computer systems, both the primary and backup systems execute the same control program to reduce the switchover time and increase dependability; this is called a hot standby arrangement.

In telecommunications, another way to reduce the risks of communication failures is through standardized interfaces. This is particularly true in the case of interactive speech communication. Standards also reduce the managerial and financial risks in joint ventures or mergers and in acquisitions of networks. Standards, however, introduce some additional risks that will be discussed later in this chapter.

Contingency planning is used for high loss events with a relatively low probability of occurrence. A set of criteria are agreed upon as the triggers to switch to a different mode of operation and declare an emergency. During network upgrades, measurements of the network congestion or response time can lead to declaration of a crisis. In the case of catastrophic situations, a disaster recovery plan would then be implemented. When a service provider uses the infrastructure of another network operator that is short on liquidity and which is being pursued by its creditors, a contingency plan is needed to prepare for an eventual shutdown of operations to minimize service disruption during the migration to a new infrastructure. In addition to the network itself of course, the plan should cover customer care procedures including constant communication with users with updates on the legal situation as well as advice or instructions on what to do. The trigger to the implement plan becomes effective when be the bankruptcy court rules to shut down the operation.

Combined Risk Avoidance and Reduction

The effects of preventive and protective actions, each alone or combined, are shown in Figure 10.2. Whenever changes in the probability or the loss function of an event changes and moves it to another quadrant, the established risk management process should be invoked.

Risk Deflection

Deflection is the transfer of the risk to another party that is more capable of facing the exposure through insurance, outsourcing, warranties, and so on. Typically, telecommunication companies have insurance policies to cover various losses or serious disruptions to cash flow.

In an insurance scheme, the project pays a premium to a company that will compensate for the damages should a risk materialize. The premium depends on the nature of the risk (root cause, probability, and loss function) and the amount of effort the project is willing to spend to avoid that event. For example, global telecommunications companies that have network operation centers in different time zones usually take insurance policies to cover the cases where the service centers are unable to operate and, in extreme cases, may move to alternative centers in another country. Likewise, corporations that have chosen to relocate their service centers should cover themselves against natural or political risks.

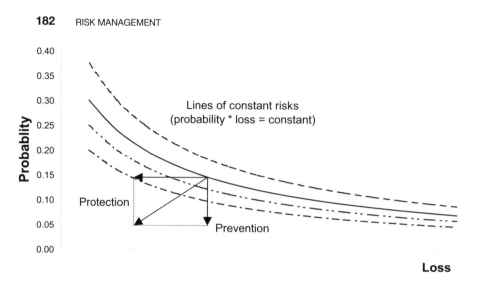

Figure 10.2 Effects of protection and prevention on project risks.

In a contracting scenario, the work is given to a supplier that is more experienced with the job or can deliver it faster, cheaper, or better. Outsourcing can provide access to skills—that is, industrial expertise that may not be available in-house or could not be acquired within the project schedule or budget.

Telecommunications companies have outsourced payroll and pension plan administration, human resources administration, and information technology (IT) support. Some such as Cable & Wireless and Brazil Telecom have even outsourced their field operation [Sirén, 2003, p. 51]. However, outsourcing key parts of the operations such as field support could increase the burden on project management, thereby risking a new type of costs and root causes for delay. In particular, it will make it more difficult to incorporate real-life experience into the planning of future projects. In the long term, it will prevent the firm from building in-house expertise to understand the limitations of the current technology and processes and hence limit the ability to forecast technology or market changes and/or to adapt to changing environment. Other outsourcing risks include breakdown in communications, difficulties in maintaining management control, susceptibility to changes in taxation, or regulations on off-shoring, and so on [Orr, 2004].

RISK FINANCING

The cost or time of finish a project is composed of three components: the cost/schedule if everything goes according to plan, the cost/schedule of doing the job if some of the anticipated risk events occur, and the cost/delay due to any event that could not have been predicted. Typically, events are divided into three general categories: *known events, known unknowns,* and *unknown unknowns.* To reduce the risk of overruns of project objectives because of the known unknowns (i.e., the identified risks) to an acceptable level, the budgeted cost or schedule is padded by adding some "contingency" to the base (or most optimistic estimate) [Desmond, 2004, pp. 67–70; Westney, 2001]. This time or money is not included in any specific project activity but is over and above the activity budget to cover a potential combination of events. In other words, contingency funding is the amount need to cover the exposure due to known unknowns.

Estimation of the contingency can rely on experience gained from past projects or through a Monte Carlo simulation and taking into account all the risk events that occur as well as their probabilities [Milosevic, 2003, pp. 300–311]. For each task that has a range of cost or duration, a probability distribution is assumed. Typically, the distribution is uniform or triangular. The output of the simulation gives a cumulative probability distribution of the final cost or schedule due to each risk category for the total project as well as for each cost and schedule element as shown in Figure 10.3. From the results, the contingency is estimated as amount needed to reach a 50% cumulative probability.

In most projects, there will be a need to cover something that was not identified as a risk at the beginning because it could not have been predicted such as unanticipated technical difficulties, labor shortages economic downturns, natural causes, loss of key people, and so on. These are the unknown unknowns. They are financed using what is called *management reserve,* which is needed in addition to the baseline budget and contingencies. This amount is not under the control of the project manager but under that of the sponsor or another higher authority that can disburse it to cover that exposure. In Figure 10.3, the management reserves are estimated as the amount needed to increase the cumulative probability from 50% to 80% [Westney, 2001]; however, a smaller number is also possible.

The key point here is that neither the contingencies nor the management reserves should be used to fund additional work due to scope changes; these are negotiated under the change control process. If at the end of the project, there is still money left in the reserves, it should be returned to the company.

RISKS IDENTIFICATION TELECOMMUNICATIONS SERVICES

We consider root causes inherent in the characteristics of the project itself, the technology, the suppliers, and the customers.

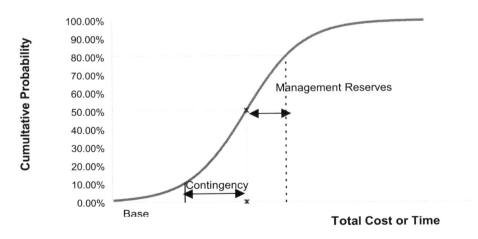

Figure 10.3   Estimated range of project cost/schedule with Monte Carlo simulation.

Project Characteristics

The risks associated with a project increase with its complexity, schedule, novelty, geography, and internal organization.

Complexity. Project complexity depends on many factors such as:

- Scope, in terms of the number of aspects that the project covers (networking technology, operations support systems and internal workflows or procedures).
- Requirements on cost, quality, or schedule and how tight they are:
  - □ Budget pressures may force the project team to eliminate a laboratory environment that could be used to analyze problems reported from the field. Therefore, any analysis will require that the failed equipment be withdrawn from the network for further testing.
  - □ Schedule pressure to accept the results for the internal testing that the supplier has conducted rather than conducting a thorough acceptance testing, which can take at least 6 months. This self-governance/self-regulation increases project risks for the following two reasons. First, the supplier does not have the time to include sufficient tests to cover all possible operational profiles, particularly for new or complicated service environments. Second, by not carrying its own testing, the project team will not acquire the understand needed to appreciate the effect of equipment limitations on the service offer and how to diagnose problems reported from the field.
- The number interfaces, particularly nonstandard interfaces, to be developed.
- Size of the project measured in the volume of orders, the number of functional groups, the intensity of coordination, the size of the budget, and so on.
- Concurrency of the tasks defined in the project plan—that is, the degree of parallelism in the project execution for faster delivery.
- Degree of interdependence with other large internal or external projects. For example, the critical path may include activities outside the project control, such as establishing local access through a different network provider.

Schedule. Long schedules increase the project exposure to risks due to (a) unexpected shortage of money, (b) changes in the technology, (c) shifts in strategy for the sponsor, supplier, or customers, or (d) changes in regulations. For example, manufacturers may discontinue their products, making their support problematic. This is of particular concern for infrastructure projects (undersea cable, satellites, etc.).

On the other hand, a short-term perspective may put the financial logic and shareholders' interests on a collision course with the project performance. The focus on short-term goals and quick results may push for speeding up the service delivery at the expense of the long-term quality of the service. Furthermore, a short-term perspective may cause pressures for shortcuts with respect to physical safety, labor laws, or environment protection. Finally, reduction of employment through layoffs or outsourcing to meet some financial objectives, as well as rapid turnover of staff and management, usually leads to a loss of institutional memory. In particular, when there is a merger or an acquisition, equipment incompatibilities may cause long-term problems.

Natural or man-made disasters can affect the schedule in many ways. At a minimum, the introduction of a new service loses its precedence ahead of that of handling the crisis and recovering from it, and a new schedule needs to be renegotiated. Also, during peak days of electric power usages, restrictions on what can be done in the network to prevent excessive power consumption and brownouts can delay the schedule. Government "emergency actions" may ask for increased security procedures that would reduce the normal productivity. A very specific case is that of preventing any electromagnetic interference with NASA (National Aeronautics and Space Administration) communications during the space shuttle launches and landings.

Novelty. The risks due to novelty depends on several factors, such as

- Type of innovation (incremental, platform, architectural, radical).
- Availability of functional and managerial skills.
- Knowledge management: knowledge acquisition, retention, development, sharing, and protection. This includes the lack of historical information on past projects. Knowledge management covers the protection of the information through a combination of security procedures and legal actions.

Geography. The geographical location of the project introduces some additional risk factors as follows:

- Networks deployed in a new country or new region may require additional testing of interfaces and then modifying the operating procedures.
- The existing infrastructure (e.g., power supplies, local access) may not be adequate for the new service.
- Infrastructure projects are susceptible to delays due to natural causes such as hurricanes, snow storms, and so on.
- Holiday schedules in different countries are asynchronous and will have to be included because they affect the availability of equipment, material, and the project budget.
- Regulations with respect to licensing requirements or equipment importation are different. In some cases, the equipment will have to be provisioned and staged in the country, rather than being performed in a central location.
- There is a jurisdiction risk in that parties to a transaction may be able to select where to initiate legal actions. For example, they might be able to find a country where a litigant can force a company to open its internal records and files.
- The particular geography of an area can cause interferences that affect wireless communications. In the United States, for example, Nextel's operation in the 800-MHz bandwidth interfered under some conditions with official communications such as emergency services.

Internal Organization. Most of the risks arise in the management of cross-functional relations and responsibilities as well as in the management of the various stakeholders. The most important sources are as follows:

- Lack of top management support or their incompetence. Symptoms include the lack of resources for the project, uncontrolled changes to the project scope, and unresolved conflicts among the functional organizations and the project organization. Another sign of this lack of support is the refusal to adopt statistical quality control to quantify risks with the result that decisions are made based on subjective criteria as well as market pressures.

- One or several of the stakeholders have a hidden agenda. This could appear as lack of coordination or conflicting objectives between the various organizations.
- Lack of commitment from all parties. For example, many qualified engineers may resist the collection of data on the overall project performance. It is up to the project manager to make the formalities of data acquisition as nonintrusive as possible and to make sure that the data collected are never used to evaluate individual team members.
- Loss or disappearance of key personnel, recruitment difficulties, low-morale, and so on.
- Lack of training of the workforce.

Technology

From a management viewpoint, the risks that the technology poses arise from applying tools and techniques that are more appropriate for phase for the technology life cycle to another.

Other sources for risks include:

- Unreliable hardware or software
- Insufficient testing
- Difficulty of downgrading the hardware or software in case of problems
- Incompatibilities with the embedded base.
- Lack of expertise with the technology
- Lack of interface standards or incorrect standards
- Health risks
- Unknown risks

Health risks affect telecommunications in the following sense. There is considerable debate on the long-term effect of nonionization radiation on users. In particular, the relation between cell-phone use and brain tumors has not been conclusively shown. Should such a link be established, we can expect a major reevaluation of the whole wireless communication industry.

Unknown risks form a class of events that appear in the field and that cannot be discovered beforehand, no matter how much effort the parties involved (service providers, vendors, customers, etc.) would spend.* Telecommunication networks of a reasonable size are extremely complex arrangements with equipment from different vendors, implementing different technologies and one or more generations of any given technology. Typically, unexpected combinations of parameters or protocols, perhaps due to misconfigurations or to some minor changes in any equipment profile in the network, at the customer premises or even in the network of another operator, may lead to considerable damage. Laboratory evaluations offer little help in discovering these "corner cases" because it

*I am grateful to Michael Recchia for reminding me to include this category of risks.

is not possible to consider the myriads of possibilities and permutations within the physical constraints on time, financial resources and human expertise. Controlled introductions would be one way to mitigate these associated risks. Another is to have a tightly controlled network maintenance program to observe the real-time performance and react to any deviation from the accepted behavior.

Supplier

Supplier's risks have been discussed in the chapter on vendor management. The main sources of risk are:

- New vendor being chosen for equipment.
- New connectivity agreements are being made. These agreements are often based on paper studies of the interfaces of the standards that are applied. Verification of the interoperability through an actual laboratory assessment is not very frequent.
- Vendor's organizational capabilities are up to par.
- Supplier fails to deliver in time and with the required quality.
- Supplier is unable to provide guidelines on how to optimize the product for the network at hand.
- In case of outsourcing, security of the information with suppliers is not guaranteed.

Customer

The degree of risk from the customer depends on the following factors:

- Criticality of the project to the customer or sponsor.
- Financial viability of the customer.
- Customer's or sponsor's unreasonable expectations on budget constraints of time, cost, and capability.
- Ability of the customer's contact in terms of skills or authority to manage the project on their side and to be ready according to the deployment plan.
- Hidden agendas/power struggles within the customer's organizations.
- The level of the end-user's support; that is, was user's input considered in defining the requirements?

RISK MITIGATION IN TELECOMMUNICATIONS SERVICES

Strategies to face risks are of prime interest to project managers and business process owners. We give in the following sections some examples of risk mitigation used by telecommunication service companies.

Risks Due to Project Characteristics

Table 10.2 lists some of the preventive and protective actions mitigate risks due to the project's characteristics.

Table 10.2   Examples for Mitigation of Risks Due to Project's Characteristics

Source of Risk	Prevention	Protection
Complexity	Reorganize the project plan to minimize overlapping activities and dependencies. Reprioritize tasks to achieve the most important objectives first. Streamline the communication among the stakeholders and team members	Engage experienced project managers. Provide training.
Schedule	Use automation to increase time saving. Postpone features that are problematic. Verify dates with vendor internal delivery milestones.	
Novelty	Evaluate vendor's internal quality data. Establish a knowledge management policy.	Training. Allow for experimentation and learning by doing. Use prototypes. Engage available expertise, if any. Coding and sharing the information gained, by having multiple teams or persons do the job (redundancy), by rearranging the work package such that the people could be rotated (partitioning).
Conflict among stakeholders	Define acceptance (success) criteria.	Control the scope.
Organization risk	Training and experimentation. Document to share information.	Rotate people. Partition the tasks among multiple teams.

Technological Risks

Some of the actions to mitigate technological risks are as follows:

- Distinguish between the needed and the desired features of the system to focus only on the needed features.
- Allow for more design iterations by using a pilot product from the supplier.
- Use standardized interfaces.
- Use a phased approach for the service introduction with various milestones for Go/No-go decisions to review the value of the project, the status of outstanding problems, and the risks.
- Deploy only hardware and software that has been thoroughly tested in the way the equipment will be used. One way to offer this is to merge multiple services on a common platform, with the added advantage of reducing the time to delivery of new services. This can be done by a staged deployment as explained in Chapter 3.

The drawback, of course, is that the challenge and cost of platform migration increases commensurably with the early gains.

- Conduct a more detailed acceptance testing for new technologies.
- Increase test coverage through automation to increase the number of test cases executed per unit of time.
- Analyze test results using the techniques of software reliability engineering to deploy a software only when the expected loss is less than the cost of continued testing, the penalty of a failure in the field, and the target probabilities of annual failure rate.
- Analyze the various defects to determine root causes and eliminate them.
- Use a controlled introduction of the equipment in the network to expand the testing operational profile beyond what is available in a laboratory environment.

Supplier's Risks

- Identify and contract with experienced suppliers, whenever possible. Start-ups may be attractive in the case of new technologies because they are more risk-oriented and can take quick decisions. Unfortunately, they typically reduce their development costs and shorten their time to market by paying less attention to quality control, which adds another level of risk.
- If during reviews of open issues logged against supplier, it appears that there is ground for concern about the viability of the supplier's approach to the project success, then some alarm should be raised to the project sponsor. Risk mitigation strategies include working with the supplier to improve the design and manufacturing processes, develop and implement design guidelines, and put more stringent quality control and audits.
- Dual sourcing may be useful in some instances. In particular, for highly reliable services and to prevent loss of service due to large-scale outages, mainly large customers require duplication of network resources. So dual sourcing can achieve this requirement since it is highly unlikely that the same problem will affect both types of equipment in the same way at the same time.
- In case of start-ups, more attention should be made to their development and quality control processes.

Customer's Risks

Table 10.3 lists some of the preventive and protective actions to mitigate customer's risks.

STANDARDIZATION AND RISK

Without standards, it would be very difficult and time-consuming to ensure smooth end-to-end communication. Yet, despite being a cornerstone of telecommunications, standards introduce risks that project managers should manage.

Standards have a life cycle of their own; that is, they change and evolve particularly in the early part of the technology cycle. Standards are modified and updated to add new

Table 10.3 Mitigation of Risks Due to Customers

Type of Risk	Prevention Strategy	Protection Strategy
Customer changes the requirements.	Use a change control strategy.	Renegotiate budget, resources and schedule.
Customer has very specific requirements that were not included in the project scope.		Use an escalation procedure to alert the management team.
Customer not ready.	Define expectations from the project management on the customer 's side. Identify project responsibilities.	Renegotiate schedule and budget.
Customer configuration incorrect or incompatible with the service.	Inform customer with configuration specifications.	Renegotiate schedule and budget.

functionalities or to remove defects (errors, ambiguities, missing details, etc). In an open environment, these changes are not under the direct control of service providers; typically, a manufacturer would like to introduce some feature to enhance their products in the marketplace. However, the impact of these additions on existing networks and services needs to be understood. This means that the telecommunication operators have to track the standards they are interested in to verify that any proposed changes and enhancements do not perturb the installed based of equipment and services.

Even if standards are stable and mature, they have to be implemented correctly. This is why to ensure high-quality services, the conformance of network equipment to the standards that they are supposed to implement needs to be assessed.

Finally, defective standards have significant costs in terms of incompatibilities, reworks, or vulnerabilities; just consider the damage that denial of service attacks have produced in IP networks. For example, link state protocols have been defined without a robust mechanism to recover from a widespread loss of topology database information or overload conditions. This has led to well-publicized service outages [Ash and Choudhury, 2004]. Also, it is well known that Multi-Protocol Label Switching (MPLS) and the associate Label Distribution Protocol (LDP) have been developed without taking into consideration the requirements for operations, administration, and maintenance (OAM) in terms of service reliability including intrinsic means for detecting and locating failures [Cavendish et al., 2004; Fang et al., 2004]. This means that network elements and services developed according to the standards have to be retrofitted to accommodate future mechanisms for failure detection.

Internal inconsistencies could eventually lead to erroneous and/or incompatible implementations, which is problematic, particularly in communication applications. Other implementation-related problems that may later hinder interoperability are missing details—due to (a) inattention to details, (b) excessive reliance on tacit knowledge or on keeping competitors at bay, (c) lack of clarity about how options should be treated during implementation, and (d) lack of information on the consequences of partial implementation of the specifications. Specifications with pseudo-code or formal languages are clear and unequivocal and help alleviate these problems. Also, there are automatic tools to check the consistency of formal descriptions and detect potential problems, such as deadlocks in communication protocols.

Some elements to consider in assessing the risks associated with an evolving standard are as follows:

- There are no—or very superficial—terms of reference for the standard, which may indicate a lack of consensus among the stakeholders.
- The standard scope is not consistent with the intended use of the standard and the phase in the life cycle of the technology. If the scope includes too many details at an early stage of the life cycle, the wrong aspects may be included.
- In the case of emerging technologies, the inclusion of many options could mean that the standard's scope is too broad or that market conditions are not well understood.
- Existence of other standard groups with similar or overlapping activities, with minimum coordination. This could be an indicator that there is (a) a competition among the standardization groups in which one will be the first to publish a specification and (b) less attention to quality of the specifications.
- No formal description.
- Inconsistency among different sections of the same document.
- No standardized tests to evaluate the conformance of an equipment to the specification.
- No pre-implementations to verify interoperability.

INNOVATION AND RISK

Data suggest that there are many strategies that are used separately or in combination to address different components of risk. Some of these strategies are used by project managers [Moynihan, 2000], while others have been typically considered in the investment community in the evaluation of their business plans [Rich and Gumpert, 1985]. The purpose of this section is to integrate all these strategies in a coherent framework by relating them to the type of innovations in telecommunication services.

Incremental Innovation

In the case of incremental innovations, the service has been developed and sold to subscribers, so the purpose of the development is to improve efficiencies of operation or expand the scale to improve growth. Thus, the various aspects of the design and project management are based on historical experience, individual or collective. Technical and managerial expertise is available in-house, through consultants or outside parties; at the same time, there are well-known ways to fill any gap in the skill set of user's and developers. During the planning phase, results from any feasibility study, historical data about similar projects, actuarial models, and the opinions of available experts help the project team capture and document sources of risks and possible mitigation strategies. In particular, the probability of the various states of nature and the payoffs are known. The technical and performance specifications are well known and are usually incorporated in the contract with penalty clauses if the reliability and performance requirements are not met. From an investment viewpoint, this kind of innovation is the most desired one (level 4 in the Rich–Gumpert evaluation system).

Some of the risks associated with incremental innovations are due to the single view of the future as a continuation of the present prevailing environment. This view can blind the project team to changes in the service environment due to leaps in technology or the changes in the business climate.

When AT&T acquired McCaw Cellular in 1994, it had to choose between enhancing a networking platform conforming to the digital standard IS-95 or adopting other technologies such as GSM or CDMA (Code division multiple access). GSM was mostly used in Europe and in the Middle East while CDMA was still unproven. Furthermore, IS-95 could evolve to IS-136 and maintain backward compatibility with AMPS (Advanced mobile phone system) the analog system that was then dominant in the United States. The decision to follow the incremental path and enhance the voice quality made it difficult for newer applications such as text messaging or transmitting pictures phone camera phones. By late 2000, AT&T Wireless had to switch to GSM at enormous costs; it had to establish a new nationwide network parallel to the one it operated while its competitors were expanding their reach and improving their service offerings. In the end, AT&T Wireless fell so far behind that its only way out was to be acquired by Cingular in 2003 [Richmar, 2004].

Architectural Innovation

Most sources of risk in architectural innovations arise from the change in suppliers, markets, customers, and so on, and the potential lack of knowledge of the new value chain. One example would be the extension of a new service offer to a new region or country. Another would be to offer new services by minor or major customizations of existing services using existing technologies—for example, selling ring tones for wireless phones. New marketing channels need to be developed, and the profiles of the new customers must be identified. Unmanaged assumptions based on past experience within the established value chain are implicit in the way organization operates and, if unchallenged, may hurt the chances of success in the new environment. In the Rich–Gumpert evaluation systems, architectural innovations are typically at level 2.

Another kind of risk in architectural innovation arises from the unintended consequences of bringing too many things together in a new combination. For example, adding Bluetooth communication to mobile phones eliminated the need for wires between the handset and the headset and made life easier for drivers. However, adding such capability changed the security equation because the communication link was not protected so that an interceptor could inject viruses to control the phone and then use it to conduct fraudulent transactions.

One of the ways for mitigating risks is to learn more about users needs through service prototypes. The need for prototyping is most needed when the customer has unrealistic expectations, the system is mission-critical or must go right the first time, it introduces major changes in the customer's workflow, and the users are inexperienced [Moynihan, 2000]. However, prototyping requires additional expenses for engineering, evaluation, and market testing to verify user's benefit and acceptance of the proposed features as well as the appropriateness of the new marketing channels. Another possible way to reduce risks is to request help from those that are familiar with that new market from scope definition throughout the execution, including training appropriate personnel. This is why collaboration with other companies is often considered even though this collaboration introduces other sets of risks, particularly if the time and cost for learning and for overcoming incompatibilities in the OSSs are not included in the budget.

Platform Innovation

The main source of risks in the case of platform innovations is that estimates for schedules and quality are based on engineering estimates and not real-life experience. Thus, despite evidence of market interest and that the new platform can enhance the service offer, there is no guarantee of either one. Also, it is not clear if the baseline design would allow future customization for architectural innovations. This why Rich and Gumpert grade this type of innovation at level 3. One way to minimize risks is to rely on available standardized interfaces or specifications or to work to establish new standards to expand the market and benefit from the collective wisdom of all other players. Another way to obtain a collective assessment of performance and risk factors is to involve technical experts as well as business process owners [Goddard and Klein, 2001].

Radical Innovation

In the case of radical innovation, the risks are associated with the steep learning curve on all aspects of the service. Subject expertise is being developed, and the problems to be faced in the field are not well understood. It is not possible to benchmark with others. Pressures to deploy the service can lead to hasty planning with no good risk assessment and with poor specifications. Some of the technical risk areas are:

1. How much research and development is required before the service is ready?
2. How much engineering is needed to design network?
3. How long would it take to certify the network elements before they can be deployed in the network?
4. How extensive would be the changes in the OSSs?
5. What is the cost of maintaining a quality operation including that of training the workforce?
6. How much changes will have to be in the workflow and working procedures?

From an organizational viewpoint, areas of focus in the risk assessment the exact nature of management support for the project including the project management process. For all these reasons, radical innovations are at level 1 of the Rich–Gumpert evaluation system. Risk mitigation strategy includes sharing risks with the suppliers, subcontractors, and even competitors such as by developing new technical standards and new business models. In addition, scoping the innovation in multiple phases using the phase-gate approach to service development helps in the control of risks.

The mitigation strategy introduces new risks, however, if the wrong standardization strategy is adopted or if the intellectual property control is either too strict or too lax for effective collaboration with other industrial players such as suppliers and competitors.

RISK MITIGATION AND ORGANIZATIONAL CULTURE

An organization's response to risk is an important factor in shaping the project's response to risk, particularly in a fluid and rapidly changing environment. One extreme is the tendency to be overly safe and exercise caution in adopting new methods and practices, while the other is a predisposition to react in an ad hoc fashion to face new challenges and opportunities. Firms motivated by safety will likely curtail the level of autonomy of the project manager. Firms thriving on risk will typically want to be pioneers but without necessarily applying the best practices in risk management. This attitude to risk can be ascertained by observing the behavior of the figures in key decision roles and not necessarily by parsing their statements. It should be noted that professional background and organizational structure play a major role in the perception and evaluation of risks.

RISK MITIGATION AND THE PROJECT MANAGER'S TOLERANCE FOR RISK

The difference between making decisions in the case of incremental innovations and the other types of innovations is that in the latter case, there are no objective probabilities available (except those derived from system test regarding reliability of the equipment as shown with the software reliability techniques). Thus, there are no meaningful assignments of probability. In these cases the assignments of probability is a group exercise in the form of a brainstorming session. However, except for the extreme cases, the ultimate decision on how to deal with risk under uncertainty is based on the subjective tolerance of the project manager to risk. There are a number of sources for biases in human assessment of information, some of which were discussed in the chapter on human resources. Another factor can be represented by a utility function, which measures the degree of satisfaction or value that the decision-maker associates with each outcome [Rubinstein, 1986, p. 235]. In formal decision theory, the utility function is used to assign numerical values to the possible outcomes of a decision model. A conservative behavior is characterized by a decreasing rate of satisfaction as the amount of expected gain increases—that is, more content to keep the available money than gamble with it to obtain a higher reward, In contrast, gambler shows an increasing rate of satisfaction when more money is at stake. Figure 10.4 illustrates the utility curves for both the conservative and gambler behaviors and the risk aversion of the conservative person.

Transposing this concept to tolerance to losses and attitude to risk, we see that for the typical exposures encountered in a project, a risk seeker project manager will prefer certain mitigation strategies (e.g, accepting the cost of increasing redundancy) while a risk aversion project manager will play it safer by partitioning the work as depicted in Figure 10.5.

SUMMARY

Risk management is a systematic way of identify and measuring risk and developing and managing options for facing those risks. Successful project risk management enhances the probability of project success by identifying critical risk areas, documenting them and including them in the over project plan is a necessary activity. Risk management is continuously linked with scope and change management throughout the life of the project. Actions to mitigate risks generate changes that, in turn, change the configuration for risks by moving the project from one quadrant to another within the bound-

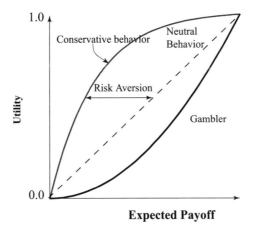

Figure 10.4   Utility functions for different behaviors.

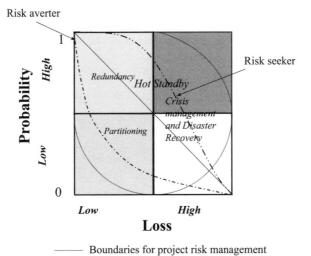

------- Boundaries for project risk management

Figure 10.5   Influence of risk preference on mitigation strategy.

aries defined by project risk management. The effect on quality, cost, and timing are updated and communicated to the stakeholders using the established communications plan so as to minimize surprises and bring all at the same level of knowledge. This allows the establishment of timely risk mitigation actions through reallocation of resources, rescheduling of activities, or scope changes. At project closure, the experience gained is to be integrated into a collective project management knowledge repository. In the future, this knowledge base can serve as a starting point for similar project for risk identification an analysis.

Standards may alleviate many of the risks associated with the technology and with vendor equipment, provided that they are well-designed and written, implemented correctly, adhered to by suppliers, and, finally, maintained to ensure their evolution.

11

SERVICE DEVELOPMENT

There are many forms of telecommunication services ranging from basic telephony using wireline or wireless access, to Internet or broadband services. Irrespective of the technology used, the establishment of these services has to go through the following steps:

- Opportunity analysis and concept definition
- Product definition and project setup
- Design and procurement
- Development, implementation, and system testing
- Service turn-up
- Commissioning and life-cycle management

The purpose of this chapter is to synthesize the various aspects of project management in an integrated plan to design, build, roll out, and operate commercial telecommunications services.

OPPORTUNITY ANALYSIS AND CONCEPT DEFINITION

The scope of this first step is to evaluate the proposed service to determine its value and the availability of the resources to deliver it in time, with the right quality and within the allocated budget. In a role model organization, the evaluation and selection process considers multiple factors such as

1. Congruence of the proposed service with the strategic objectives of the company and its consistency with its image in the marketplace

2. Service objectives and position with respect to other offers
3. Financial objectives in terms of revenues and profitability
4. Internal capabilities including the skills and motivation available workforce
5. Regulatory environment
6. The various risks that the project could face, include changing regulations, technology difficulty, market shifts, and so on.

The output from such an evaluation includes an initial business case, a preliminary product description, and a general impact statement on the existing network infrastructure as well as other services in the marketplace.

PRODUCT DEFINITION AND PROJECT SETUP

The activities in these phase turn around refinements of the outputs from the previous phase. More specifically, the main tasks are:

- Clarification of the sponsor's objectives and the project scope.
- Definition of the project charter, project plan, and resources.
- Specification of the service in terms of the technology and the market sector to be served. In general, the marketing analysis of a sustaining innovation can rely on the customary tools; for a disruptive innovation, however, the definition has to be tentative and must include agreed criteria to track the environment and determine when corrective actions may be warranted.
- Estimation and justification of the investment required through a business case.

Clearly, the intent is to explore the extent of management's commitment to the stated objectives. Typically, the introduction of new telecommunications services go through three successive waves: first, to establish market presence; next, to increase market penetration; and, finally, to recover the investment and generate profits. Management has to define where and when they plan to be. If the product is brought to the market in the shortest time, how much reduction in the services features or quality are they willing to allow? Or how much additional funding will they make available to the project team? In the ramp-up phase, where are they aiming the expansion to be and at level of financing? Finally, when the service is expected to be profitable, what is the expected life of the service?

A project charter documents which of these goals and parameters guide the current endeavor in terms of the business and political drivers, the various stakeholders, their roles and responsibilities, the metrics for success in terms client satisfaction, and the triple constraints performance. This charter becomes the guide for the project management team as they scope and assemble the project team. With the project management structure established, perhaps in the form of a program office, the next step is to define the project plan. That plan deals with the various aspects of the project (scope, resources, communication, quality management, etc.) including progress indicators, measurements to assess the client/customer satisfaction indicators, and criteria to track the ongoing project viability. A more precise definition of the opportunities and threats takes into account an updated view of the competition, the regulatory environment, technology trends, and market shifts.

DESIGN AND PROCUREMENT

Typically, planning of all projects covers the following aspects:

1. Definition of an architecture to meet the project objectives
2. Supplier management
3. Technical definition of the service
4. Operations support systems (OSS)
5. Disaster recovery
6. Communications plan

In the case of new service introductions (i.e., nonincremental innovations), additional aspects include marketing and sales plan and regulatory plan.

Architecture Design

The service architecture has to cover the geographic span of the service, the performance characteristics, and the service availability.* Service availability impacts the design of the overall topology, to maintain network connectivity for customers during traffic rerouting over alternative routes [Bhandari, 1999]. It also imposes requirements on the redundancy of equipment and diversity of routes.

In 1988, a fire in the Hinsdale central office caused the loss of service to more than one million subscribers in Chicago's western and southwestern suburbs for about one month. In the aftermath of that event, architectural decisions were made to make central offices more robust to single failures through redundancy [Wrobel, 1990]. More than three decades later, the damage to a Verizon Commmunications switching hub that abutted the World Trade Center during the September 11, 2001 attack on Manhattan left about 300,000 voice lines and 2.5 million data circuits without service; three months later, an estimated 57,000 lines had not been restored [*Wall Street Journal,* 2001]. This event has stimulated many institutions to review the backup plans of their private networks with an eye on diversity—with at least two carriers having two independent routes to two different hubs—to retain connectivity should another worst-case scenario turn into a reality.

Service protection can be done either by partitioning or by duplication.

Redundancy covers components and cards in network elements, backup circuits, and trunks for rerouting traffic when isolated network components fail. Diversity of paths can be reached through dual local access (*dual homing*) to the central offices or point-of-presences in critical sites as required.

Protection by partitioning gives the ability to recover from failures in certain nodes or regions by dividing the network in two parts, each carrying one half of the traffic. This allows graceful degradation of the service because the probability that both sections will be affected will be less. Figure 11.1 shows such an arrangement in a corporate network.

---

*Availability of service is the property that makes that service dependable and satisfactory for users as described in ITU-T X.137, X.140, and X.800. Diversity is the property that the same origination and the same termination are connected by different and separate paths (M.495, M.3208.1). The reliability gives the time-dependent probability that an item (hardware or software) is capable of performing a function for a given amount of time (E.800.) Finally, redundancy is the property of duplication due to the existence of more than one means to perform the required function (E.800).

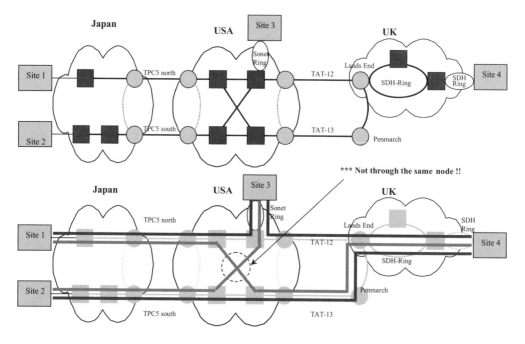

Figure 11.1 Protection of network traffic by partitioning.

Protection against outage by duplication is based on physical route diversity and more robust network topology to re-route around major outages. All resources are duplicated: the access lines, the switches, transmission links, the points of presence (POP) of the carrier, and so on. This allows switching the traffic after a failure from the damaged primary network to the backup network. It is even possible to have two different switches in each network, or from two different suppliers to avoid any interruption due to failures specific to one machine or due to the lack of hardware. The backup network is used on a hot stand-by basis to reduce the switchover time whenever the first network is not operational. This diversity, of course, would not be cheap. On the average, the cost of a truly duplicate network will be have to include the cost of each carrier network as well as the cost of the equipment that detect the failure and initiate the switch over. Figure 11.2 illustrates the protection of an enterprise network through duplication of resources.

It should be noted, however, that true diversity is also difficult in cities where there is strict control over rights of way and only one company is allowed access to a building. In Manhattan, for example, the rights of way are owned by one of two companies: Empire City Subway Co., a Verizon subsidiary; and Con Ed Communications, Inc., a unit of the power company Consolidated Edison, Inc.

BITS, a sister organization to the Financial Services Roundtable, has published a guide to help business managers and continuity planners analyze and manage risks associated with critical telecommunication services [BITS, 2004].

Supplier Management

A supplier management process is established to select the suppliers and to define the evaluation and acceptance criteria in the purchase contract (e.g., verify that there are no

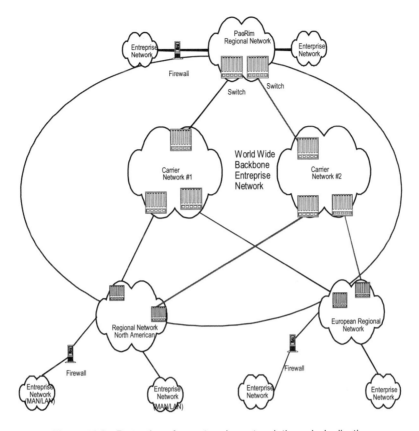

Figure 11.2   Protection of an enterprise network through duplication.

critical or major defects, pass a 48-hour stress test, etc.). Ways to collaborate with suppliers and exchange information (perhaps through an extranet) must be devised to ensure timely execution and lower implementation costs. Procedures and systems must be available to maintain all purchase orders, service orders, vendor contracts, and leasing agreements that would be needed for disaster recovery and restoration.

Procedures to return defective items, sometimes called return maintenance authorization (RMA), define how to return failed hardware component failures and exchange them with new ones on a 7 × 24 basis. The main components of such a procedure are:

1. An accurate and up-to-date inventory of the maintenance spares
2. An up-to-date list of the hardware certified to be compatible with the deployed configuration
3. Test procedures for the suspect hardware
4. Contact points and a return address to return the failed hardware and replace them with items from the certified
5. Defined shipping procedures and documents (including customs document in case of foreign locates) to have the replacement shipped to replenish the local depot
6. Procedures to track the status of the hardware in shipment and in storage so that the

procurement department is aware at all times of the location and the status of all parts

7. Tests to verify the revision levels of the replacement hardware to ensure that they are on the certified list

Technical Definition of the Service

The technical service description documents the planned mix of services; their performance and their availability over time are described. The description includes anticipated traffic volumes, their geographic and temporal distribution, the billing elements per service, the geography of service availability, the customers' profiles, and so on. Other criteria include the quality of the service in terms of reliability, availability, service availability, and lead times. The technical service description identifies the various national and international regulations and standards to comply with. It defines site requirements and specifies the compatibility with other networks and services.

Ideally, the design will be scalable to accommodate future volume requirements, extendible to support future new types of services, and evolutionary to integrate future switching technologies.

Site Selection. Two types of sites are needed: sites to host network equipment [points of presence (POPs)] and/or to stock spare parts and work centers.

The sites that host equipment can be owned or rented through telehousing agreements. Site specifications cover floor space, rooms, buildings, electrical power (uninterruptible power supplies including battery and diesel generator), connector types to the equipment, synchronization and clock feeds, human resources, security procedures, and so on. Safety (e.g., fire resistance, electric insulation, earthquake resistance, etc.) and environmental concerns will be based on the existing regulations as well as the requirements of the potential customers. If the sites are already used, an inspection visit gives the opportunity to discuss with the site supervisors and the technicians to learn about the existing conditions and gets their approval for the project. Based on this visit, construction and refurbishment plans may be made.

The work center is the physical site where people provide various roles that are usually classified under customer care or network care.

Service Operations Technical Plan (SOTP)

A service operations technical plan (SOTP) describes the various processes and systems for sales, ordering, provisioning, maintenance, and capacity management to support the quality objectives defined in the project scope. The SOTP covers various processes ranging from business management to network element management, the roles and responsibilities of those involved of each functional area, the documents or forms (whether paper or electronic), and the systems or tools used, that is, the Operation Support Systems (OSS).

Figure 11.3 gives a three-dimensional representation of the relationship of the support systems with organizations and processes. This representation, due to Steve Pollak, is inspired by the Telecommunication Management Network (TMN) methodology of the International Telecommunication Union—Telecommunication Standardization Sector (ITU-T) [ex-CCITT (Comité Consultatif International Téléphonique et Télégraphique] as specified in ITU-T Recommendation M.3200.

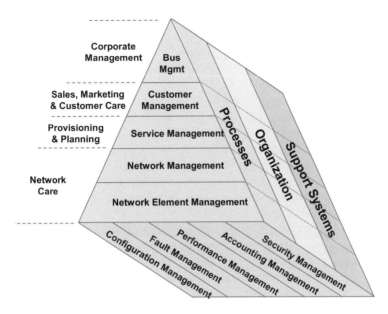

Figure 11.3 Management systems model developed by Steve Pollak.

Support Processes. The transmission infrastructure will also include the data communication network (DCN) that is used by the OSSs.

The main processes involved in service development are:

1. Pre-sales and sales
2. Service ordering
3. Provisioning of the network configuration according to the terms of contract and the engineering rules of the network
4. Operations, Administration, and Maintenance (OAM)
5. Accounting and billing

All these processes must be able to cope with future business evolution (more customers, changes in traffic volumes and patterns, etc.) to ensure continued operation at the agreed levels of performance as the service expands.

Acquisition and Sales. The function addresses all the activities that precede sales. It includes lead generation, prequalification, funnel analysis, proposal development, network configuration, and pricing and contract development.

The pre-sales/sales process begins with an initial customer contact or inquiry to an account executive and concludes with a signed contract and the hand-off to the ordering and provisioning organizations. The sales organization depends on several tools for configuration and pricing. A network design is the basis for discussions with the capacity management organization as well with other bandwidth providers for access and transport (back-haul), to get pricing quotes and discuss availability. If the customer premise equipment (CPE) is not in the list, additional testing may have to be conducted with a loaner from the customer or the CPE vendor some time before the service date (around 4–6 weeks). It is seen that the determination a network activation date may require several iterations among the customer, the sales team, the engineering group, and the capacity management organization within the telecommunications provider as well as within outside bandwidth providers.

Service Ordering. Service ordering begins after a sale has been made and a contract signed. This process requires consolidation of customer and contract information to issue the different elements needed for fulfilling the service order and feeding the downstream processes that provision, maintain, and bill for the services offered. Such elements include customer premises information, location, contact, billing information, and, if necessary, the type of CPE. Based on this information, it would be decided if CPE compatibility with the network needs to be tested and how access to the customer premises can be made. A service ordering tracking and archival system should consolidate the information regarding all pending orders to track their status and notify customers and other functional entities whenever needed.

The service ordering process should consider standard requests as well as special requests, such as expedited requests.

The service ordering processing interfaces with capacity management system to verify the availability inventory of ports or switch terminations. The capacity planning organization is responsible for ongoing network extensions to meet business forecasts in a timely fashion based upon traffic projections and expected contracts as estimated by the marketing department. Clearly, the success of the service ordering and capacity management depend on good record-keeping to have an up-to-date and accurate inventory.

Provisioning. Provisioning is the set processes used to configure the network resources to fulfill the service requirements according to the network engineering rules. It includes entering the necessary parameters on various equipment as well as backend support including those for inventory management and network monitoring. Good record-keeping and accuracy are essential; ideally, human intervention should be minimized to reduce errors. For example, incorrect labeling of circuits can mislead technicians and cause them to bring down live circuits; this may also delay the diagnostics of faulty elements.

Accounting/Billing. Accounting/billing management systems have two components: business rules and interfaces to other management systems such as the network element management systems and the account receivable systems.

The business rules determine the way a bill is formed from the records collected from the network elements, taking into consideration the account rates, any promotion or term plans, and quantity discounts. Thus, the correct configuration of the billing system depends on information provided from the upstream sales and provisioning processes. The billing elements of a voice call have traditionally been *call detail records* (CDRs) and the *tariff table,* which takes into account the business rules. A rating engine combines these two inputs to produce the charges. The output of the rating engine is then converted into a bill with a specific format that is sent to the user either electronically or in paper format. This can be done on a periodic basis or, in the case of high-end systems, in real time. Upon receipt of the payment, it is reconciled with the account and sent for compensation through the banking system. If electronic billing and payment methods are used, the security of payments and privacy of data must be ensured.

The process becomes more complicated when used for private networks that span several countries depending on whether the billing is centralized or decentralized and the currency used for invoicing.

When market pressures force an early service introduction before all OSSs have been readied, some manual intervention may be unavoidable. Typically, it is the automatic billing capability that is postponed.

Operations, Administration, and Maintenance (OA&M). The processes for Operations, Administration, and Maintenance (OA&M) concern the day-to-day activities to run the service and meet the guarantees of the service level agreements (SLAs) such as performance and security. The technical aspect of the OA&M relate to the capability to diagnose the problems and to localize faults, which depends on the back office systems for testing and archival, the expertise of the workforce, the capabilities of the technology itself, the properties of the equipment, and the level of vendor engagement.

Finally, security management includes, but is not limited to, physical and logical access control to central offices or the network(s), data integrity, data protection, customer information protection, and so on.

The OSS must meet the needs of two different sets of customers, internal and external. Internal customers are typically interested in interrogating the billing system to evaluate the current usage of the service taken as a whole and for some specific customers or customer profiles so that, when correlated with other data, they can improve the positioning of the service by bundling it with other services or by defining new discount structures, and so on. Traffic records are also needed to validate the network engineering rules and to model future trends. External customers are looking for detailed and transparent bills for managerial control, including verification of the bills accuracy. A multinational company may also be interested in a single bill covering the whole of its world operations. Others may want bills sent to individual management units in specific formats or languages.

Maintenance Processes. Network monitoring aims at reducing the risk of outages and maintaining the performance of the various subnetworks providing the end-to-end connection. A typical in-service monitoring plan defines:

1. The type of the data to be collected to characterize failures or intrusions
2. The methods used for collecting the data—for example, intrusive or nonintrusive testing
3. The procedures to be used for processing and analyzing the data
4. The network maintenance organization, procedures, and tools

There are two kinds of maintenance functions: preventive or scheduled and corrective or unscheduled.

Scheduled Maintenance. The scheduled (routine or preventive) maintenance depends on scheduled outages to fix software and hardware problems or to perform upgrades. This type of maintenance is reserved for off-peak hours to minimize customer inconvenience. While in public voice networks scheduled maintenance is typically transparent to the users, an accepted practice in data networks is to notify customers and users 4–8 weeks ahead of the planned outage to give them a change to rework their schedules or propose different dates.

Unscheduled Maintenance. Unscheduled (corrective) maintenance is a response to problems that affect the service quality such as trunk congestion, blockage or misrouting of traffic, call establishment failures, transmission cuts, and so on, as indicated by network alarms or by customer trouble reports. Trouble response has two aspects: administrative and technical. The first deals with the reorganization of the workflow for maintenance and response to failures, for recording the details of the trouble reports, and for interface with customers to notify on repairs, planned outages, and so on.

The maintenance process consists of the following steps: (1) fault identification, (2) network event evaluation, (3) creation of a trouble report, (4) trouble isolation and diagnosis, (5) service restoration and network repair, and (6) test and turn-up.

Many of these steps require manual intervention, although it is possible to automate some of them by accumulating knowledge on the type of faults and their root causes to allow early identification of recurring failures or chronic trends. Such automation requires filtering and correlation of alarms so as to relate multiple alarms to the root cause failure. Trouble reports may also come from users' complaints.

Trouble Reports (Tickets). A trouble report or ticket contains several pieces of data such as the date and time of the problem, the type of the problem, the symptoms of the failure, who is currently on the issue, and the status of the resolution. Once a ticket is logged, it is processed according to its priority. Once the repair is made and verified, the ticket is closed and the customer is notified with a completion report. Otherwise, a report on the progress is provided to the customer on a periodic basis. The incident ends when the trouble is repaired, verified, and accepted by the customer with a completion number.

Escalation Procedures. Escalation procedures are invoked when the fault isolation and/or repair take longer than what is specified in the service agreement or when the resolution of the problem requires resources beyond local control. One possible escalation hierarchy is shown in the Table 11.1 for two categories of service: an average category and a high-quality category.

The performance of the corrective maintenance can be evaluated through outage reports.

A catastrophic failure in a data network is defined as an event such that (1) the service becomes unavailable to 25% or more to customers on a node or, (2) 10% or more to the customers serviced across the network lose the service, or (3) no resolution to the problem has been identified within two hours. An extreme case is what is called network "melt-

Table 11.1 Escalation Levels

Escalation Level	Maximum Time (hours) Before Escalation	
	Average Service	High-Quality Service
1st	2	0.5
2nd	4	1
3rd	6	2
4th	—	4

down" when an anomaly propagates from node to node, causing an overload condition, usually due to an excessive number of network control messages that overwhelm certain processors, perturbing the service within a large geographic area. Networks that use link state protocols are more exposed to this problem because each node has to inform other switches whenever there is a change in the trunk status. Corrective procedures must address the following aspects:

1. Immediate relief activities to reduce the impact of the network outage
2. Isolation of the failed region
3. Control of the situation
4. Returning the network to a stable situation
5. Short-term monitoring to prevent the problem from reappearing
6. Long-term plans to avoid future problems

Disaster Recovery. Many telecommunication outages are due to reasons beyond the control of the service or network providers. Careless digging along major fiber-optic routes by construction companies provoke cable cuts. A major power outage in Northeastern United States and Canada in the summer of 2003 knocked out wireless communications. The damage to public networks subsequent to the disaster of September 11, 2001 in Manhattan affected telephone central offices. In 1988, a fire in the Hinsdale, Illinois Central Office knocked out 43,000 lines and affected more than 500,000 subscribers. For all these reasons, a recovery management plan must be ready to provide a blueprint for a quick reorganization and coordination of activities under emergencies whenever widespread networking disruptions take place due to equipment failures, fires, floods, earthquakes, or national emergencies. In general, the plan has to address the following aspects (regulations may also mandate additional steps):

1. Relief activities at the site of the disaster event to reduce the impact of the disaster on the employees, their families, and communities
2. Recovery activities to provide a temporary, "survival" level of service by reestablishing the most critical business processes and functions in one or more sites other than the affected site
3. Restoration activities to fully restore the business functions or processes to their normal capabilities through rerouting of traffic, rebuilding of a failed site, movement to a new site, absorption of the business function into existing facilities, and so on
4. Internal and external communication plan to give updates on the status of the activities

Planning Steps. Planning for disaster recovery and service restoration involves several steps:

1. Identify critical elements that can cause an outage.
2. Utilize extensive analysis, modeling, and simulation to understand (a) the various combinations of factors that can cause disruptions and (b) how the disruptions could manifest themselves and ways to control the situation. Risk assessment and mitigation are done at a network element level, a software release level, and a soft-

ware/hardware combination level because each combination may have its own peculiarities.

3. Plan and prepare the recovery including a management escalation process.
4. Document the recovery and restoration plans.
5. Approve and distribute the plan and integrate with normal operating procedures.
6. Execute the plan through paper review of the disaster recovery plan, walk-through, or simulation exercises as well as during a actual event. Record the exercise history.

   - Walk through: This is an exercise in which the team talks through what members will do in various scenarios so as to check the logic of the documented recovery plan.
   - Simulated Failure: This is an actual execution of one or more component of the plan to verify its validity—for example, to check the contact list, the recovery procedures, and so on.

7. Establish change control procedures to review and update the plan.
8. Strive for continuous improvement based on simulated or actual disaster recovery occurrences. For example, review the validity of the network engineering parameters, assess the need to upgrade to higher-speed processors, or develop network management tools to anticipate buildup of message volumes and allow active intervention prior the buildup of service affecting loads.

Content of the Plan. The recovery plan coordinates activities at each recovery site and, when applicable, among recovery sites to meet the objectives. This will be done by dividing the recovery procedures into manageable units and then defining the procedures, sequence of activities, potential decision points that may be encountered during the recovery, the criteria for reaching a decision at those decision points, and a chain of command to face unanticipated events. The key elements of that plan are:

1. The recovery team
2. The milestone events for triggering the execution of the plan—in particular, the triggering circumstances and who can invoke the plan
3. The order and flow of activities that must occur at the designated recovery site(s) for partial restoration of the service
4. The decision chain of command in case of unanticipated events or problems during the recovery process (e.g., unexpected response in the network)
5. The sequence of tasks and related procedures for the recovery, including the coordination among the various teams and their dependencies

Customer Network Management. New networks may provide their customers with the capabilities of administering their private networks in terms of changing the configuration parameters, placing orders, monitoring the progress of installations or the traffic flow, receiving alarm notifications, and accessing traffic statistics and reports. They may be able to exchange accounting information with the network operator to verify the accuracy of the billing information. One possible architecture for customer network management (CNM) is defined in ITU-T Recommendations X.160, X.161, and X.162. Figure 11.4 illustrates how an authorized user at the customer's site can access the relevant performance data, provided that a single network provider is responsible for managing the private network.

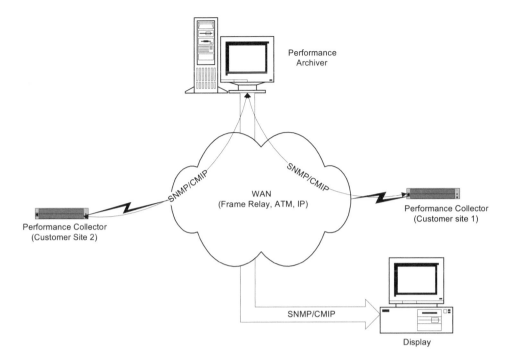

Figure 11.4 Data collection for customer network management.

However, when several network providers participate in the end-to-end service delivery, because the sites are in various geographic locations or countries not served by the same network providers, all these providers have to agree on where to collect the data and how to share them over several administrative domains. Although there is a standard way on how to exchange trouble tickets for leased lines using ITU-T Recommendation X.790 as a basis, there is no a standard way for other network or technologies.

DEVELOPMENT

Members of the service development organization provide the engineering and design rules to the organization for the provisioning and the ongoing maintenance of the service. The detailed implementation plan is specified during the high-level and detailed design phases. These rules cover the procedures for network care, customer care, and other operational support procedures. This includes securing the network elements and network element management systems, testing them for acceptance, and shipping them to the sites.

Equipment Handoff

At equipment delivery, a vendor will have to provide the following items:

1. Release notes specifying the defects that have been fixed since the previous release,

known bugs that remain, exception conditions, and the number of noncommented lines of code changes. The latter figure is used to get a rough estimate of the software reliability.

2. Software installation and back-out procedures

3. Upgrade/downgrade procedures

The acceptance report indicates the hardware and software versions accepted, the compatibility with other systems in network, and any known problems that were uncovered during the testing and have not yet been repaired and their impacts on the system performance.

System and Integration Testing

Delivered components and equipment will go through system and integration tests. The goal is to validate the network engineering rules, particularly to optimize cost performance, and to verify that the equipment can deliver the service with the specified quality requirements. The tests include:

- Functional tests for the product features
- System verification test plans and results
- Integration test plans and results
- Regression test results (in case of testing new versions of the same software)
- Performance characterization test plans and results, particularly 7-day "soak tests" and 48-hour "stress tests." Soak tests are those where the equipment sustain data transfer at 80% of the capacity without planned or unplanned downtime. Stress tests verify that the equipment can sustain data transfer rates at full capacity without causing problems.

Integration testing verifies that the new equipment can coexist with other existing equipment in the network without degrading the quality of the service. This may involve interoperability with the CPE, which depends on cooperation of vendors and customers to perform the necessary tests. Results from all these tests are used to fine-tune the configuration of network elements.

The best situation is to have three test beds, one for functional testing of the equipment (i.e., to test new hardware or new software releases), one for integration testing with other OSSs, and one for the analysis of problems that may appear in the field. The alternative is to use the same test bed for both, which would require periodic upgrades and downgrades, thereby delaying all other activities.

All scripts that will be used to monitor the network and to collect accounting data during operation must be tested.

When problems are detected during the testing, it should be identified what problems need to be fixed and which ones can be postponed to a later date. Negotiations with the supplier define how to treat these problems.

The acceptance report indicates the hadware and software versions accepted, their compatibility with other systems already deployed in the network and any known problems that were deferred and ways to avoid them.

Based on the test results, the final documentation is prepared regarding the network operational procedures and the way the service will be positioned, sold, ordered, provisioned, installed billed, and maintained.

Network Operations Center (NOC)

The network operations center (NOC) is the single point of contact for the operations, management, and administration of the network infrastructure. Among the many responsibilities of the on-site workforce (OSWF) in the NOC are network monitoring and preventative and corrective maintenance. Since in a typical telecommunications operation the OSWF is responsible for the day-to-day operation of many service offers, the guiding philosophy is to have sufficient coverage at an economic cost by eliminating duplications and implementing consistent procedures across services. It should be noted, however, that to maintain service availability under the worst conditions, some redundancy is unavoidable.

Human Resources. The human resources plan contains procedures to attract, retain, reward, and develop qualified people in the work centers. It should cover centrally located and remote staff, especially in foreign countries, as well as the requirements on subcontractors. Training ensures the transfer of knowledge to the various levels of network support. For example, training of Tier 1 and Tier 2 workforce allows them to fill trouble tickets correctly and to repair first-order troubles.

Return Maintenance Authorization (RMA). A main design issue in the RMA process is whether the replenishment parts are stored in a central location or on-site. In either case, several additional processes need to be established:

1. Maintain an up-to-date list of hardware (e.g., a list per manufacturer or per site, etc.).
2. Define the RMA treatment per category. Critical RMAs relate to urgent replacements of parts that are essential for network operation, while the maintenance RMAs are routine updates to replace discontinued versions with current versions.
3. Give the OSWF an accurate readout of the status of the parts.
4. Verify that spares are of the correct revision and are in working order.
5. Measure the manufacturers' performance (i.e., the time they take to dispatch parts to replenish the store) to assist in the management of the supplier's performance and obtaining refunds.

Customer Care

Customer care is the management of all customer-facing functions that support the service offers. Typically, a call-center provides a single point of contact for queries, new service orders, requests for modification, trouble reports, and update information on repair and service restoration. Thus, customer care is another entry point to identify network troubles.

Customer care is an important differentiating and a critical success factor to succeed as operator in the deregulated and competitive telecommunications services industry. A dual constraint weighs on the customer care organization: in terms of the volume of inquiries and the quality of the responses. Typical quality measures monitor the characteristics of the service queue such as arrival rate, wait time, percentage of callers who desist, percentage of unanswered requests, average time for an answer, and so on. Using control charts, it is possible to determine if the service delivery process is under statistical control. Once it is stable, then it is possible of ways to improve the delivery through incremental changes to reduce the cost or improve the efficiency.

In the case of incremental service innovations, customer care can be optimized with automation including scripts that guide the operators' interventions. In the case of other innovation types, some process must be established to gain knowledge from experience before any attempts at improvements are made. In a tiered support operation, customer requests are routed depending on the type of the inquiry, the severity or urgency of the problem, and the degree of technical expertise needed for response. Technical experts from the functional organizations or from the vendors may be called upon to address unusual, complicated, or critical problems.

SERVICE TURN-UP

Service turn-up may include up to five phases: installing the equipment, on-site testing of equipment, pilot testing, controlled introduction, and general availability.

Installing the Equipment

The sites to host equipment are prepared and the equipment is installed. Tasks include mounting the equipment, cabling, fanning and connecting the equipment, powering it, and test and turn-up. In some cases, right of ways must be secured.

In-Field Tests

Switch validation testing (SVT) consists of turning up the switch and running the diagnostics that the supplier has provided.

The network validation test (NVT) ascertains that the installed element can communicate with the rest of the network elements and network entities.

The scope of the service validation testing (SVT) is the end-to-end service definition including the transmission path (equipment on each side and the repeaters on the connection), the network response time, round-trip delays, and so on. All these tests are done without any end-user's traffic. Alarm and network management tests cover the operations, administration, and maintenance of each end to verify that network alarms are correlated with end-user provisioning data.

The operation readiness testing (ORT) concerns the readiness of the whole organization before service introduction, including all processes and the on-site workforce (OSWF) by going through a simulated order from one end of the process to the other (including cutting a bill). The purpose of the ORT is to exercise the linkages among the various functional processes including sales, ordering, testing, billing, maintenance, and return material authorization. For example, calls can be placed to the customer care system to test responses to failures.

The ORT entrance criteria are as follows:

1. Methods and procedures have been developed.
2. Administrative processes have been defined, including any necessary manual work-around.
3. Work center personnel have been trained.
4. All support systems are functional (exceptions should be noted).

One important aspect to test is the accuracy of the circuit labeling schemes, particularly if there are several operators involved to make sure that there is not a mismatch between the reference numbers recorded in the various network inventories. Mismatches would cause confusion in case of trouble or would prevent fault localization. Another important aspect is to verify that the RMA procedures identify uncertified hardware before its deployment in the network.

Any issues raised during the ORT must have a plan for resolution.

Pilot Trials

The purpose of pilot trials is to evaluate the service delivery capabilities of the infrastructure, the supporting organizations, and the interconnections to external networks using objective measures on the performance as well as subjective measures from user's feedback. The user-oriented metrics differ depending on the type of traffic (voice, facsimile, data, video, etc.). Correct assessment may require the development of a service failure impact matrix to evaluate the impact of service failure on the customer services.

The extent of testing depends on the innovation. For incremental innovations, controlled introduction may suffice; for other innovations, however, more extensive trials are warranted. *Alpha trials* focus on simulated customer settings to test the preproduction systems. *Beta trials,* in contrast, recruit pilot customers—that is, those ready to participate with real applications without expecting the service levels that will be provided under real operation. Those who agree to participate in such trials may want to give their input to improve the service or would like to use the offer in return of a reduced rate or a free service. Finally, pilot trials for nonincremental innovations showcasing a product may be an important part of the launch effort.

Evaluation of the pilot will cover at least the following aspects:

1. Performance of the network with respect to the design requirements.
2. Effectiveness of the service delivery process in terms of on-time provisioning.
3. Effectiveness of the preventive maintenance and the corrective maintenance processes.
4. Effectiveness of the customer care organization in answering inquiries and complaints.
5. Accuracy of the billing systems
6. Effectiveness of the education and training of the workforce OSWF.
7. Responsiveness of the equipment supplier to field problems and the effectiveness of the vendor interfaces such as RMA for replacing defective cards.

When the pilot test ends and the equipment is accepted, the life-cycle management organization will take responsibility of the network operation, provided that whatever problems are discovered do not prevent the deployment of the equipment. Results for the pilot testing may be shared with the vendor if the equipment will have to be improved.

Controlled Introduction

As explained in Chapters 8 and 10, laboratory results give a good indication on field performance, provided that the test profiles are similar to the operational profile. A controlled introduction is a way to test this hypothesis by deploying the equipment in selected sites and observing the behavior with real traffic before deciding on further deployment. If the patterns of traffic are close to the profile used in the system test and no problems are discovered then the service can be comissioned for general availability. If problems arise or if unexpected traffic patterns are observed, then further evaluation is needed.

Management of the Controlled Introduction.   During the life cycle of the project of the service, there will be several software and hardware upgrades. The controlled introduction will be sued to develop a process to management the technical and managerial aspects of these upgrades. The technical aspects include the following:

1. The entrance criteria for planning of an upgrade
2. Upgrade and downgrade procedures (in case there is a problem with the new version of the software)
3. Scripts to monitor the status of the network to make sure that no problems occur
4. Methods to baseline the entire network before the upgrade to help in monitoring the network status during the upgrade
5. Tests circuits to monitor the status of the network

The management aspects include the following:

1. The command structure for decision during an upgrade. This includes the establishment of a command center (a "war room") and a list of technical and managerial contacts.
2. A crisis management or contingency plan, in case there is a problem during the upgrade—for example, a large deviation from the baseline status or an increase in the number of reported customer troubles.
3. Training plan for all personnel involved in the upgrade.
4. Communication plan to notify customers with time frame of upgrade, its impact on their traffic, and how they will be informed on the status of the upgrade.
5. The post-upgrade hand-over procedure to the line organization for life-cycle management.
6. Role of the on-site vendor technical support team.
7. Prepare system to record detailed description of events during the upgrade for future review.

Crisis Management Plan. A crisis management should be in place in case there is a problem during the upgrade. The plan defines the criteria to trigger the plan and declare an emergency, the authorities that could declare or call-off the crisis status, and what should happen when an emergency is declared (procedures, contacts, etc.). Some of the steps include: the formation of teams for 24 × 7 coverage, where they will be located, their roles and responsibilities, the communication plan; what objective criteria will be used for aborting an upgrade or for a downgrading the network to a previous release; and, finally, the exit criteria from the emergency situation.

For example, one reaction would be to form three distinct teams consisting of personnel from the network operator and the equipment manufacturer: an upgrade monitoring team, an analysis team, and a management team. Once an emergency is declared, the upgrade team will supply information to the analysis team. The analysis team will try to understand the origin of the problem, reproduce them in a laboratory environment, and make a recommendation to the manager on duty. Based on the available information, the management team can decide on the course of action, such as continuing the upgrade, selectively downgrading some nodes, or aborting the upgrade and backing-out completely.

Time is of critical consideration before declaring an emergency. Accordingly, determination of whether the source of the anomalous behavior is hardware or software needs to be accomplished in 30 minutes during the upgrade and 2 hours for post-upgrade. Furthermore, determination of whether there is a workaround must be done in 30 minutes during the upgrade or 2 hours for post-upgrade. Based on the available information, the management team can decide on the course of action—for example,

1. Abort the upgrade and back-out completely.
2. Selectively downgrade of some nodes.
3. Continue the upgrade if there are known workarounds, and so on.

The decision process and a proposed time scale is shown in Figure 11.5.

In any event, all parties must be invited to contribute to the development of the plan so that it represents the collective view of what has to be done. This is an iterative process that may take some time in the first round; however, it is essential for a switch transition from the normal structure to the structure needed to manage the crisis efficiently.

Final Marketing and Sales Plans for General Availability

The market support plan outlines (a) the sales and revenue forecasts for the product overall and for each market into which it will be sold, (b) the pricing and introductory promotions for each market, (c) special conditions and rules for provisioning the service, and (d) the training requirements for field and branch office support.

The sales plan defines the launch data and the specific launch plans for each sales channel as well as the requirements for each channel support. For each channel, it establishes sale targets and revenues per channel as well as a list of contacts.

An advertisement plan defines the various promotions and public relations activities such as press releases and publicity campaigns.

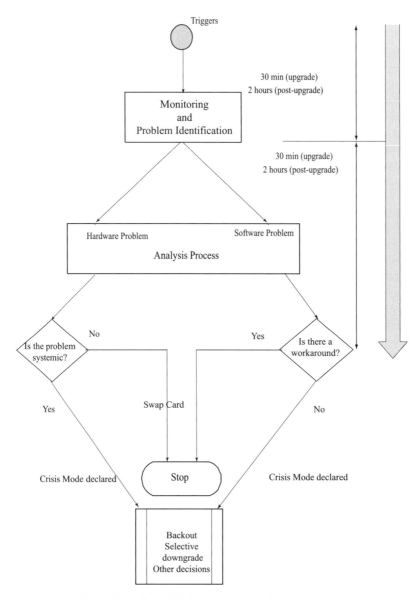

Figure 11.5 Potential decision process during an upgrade.

COMMISSIONING AND LIFE-CYCLE MANAGEMENT

To manage the transition of life-cycle operations to the line organizations effectively, representatives from each functional area must take ownership of their activities. The decision on where to launch a service and in what order depends on many factors such as regulatory conditions, vendor support capabilities, market conditions, competitors' moves, readiness of the workforce, readiness of the OSS, the sales and distribution channels, and so on. When the service general availability proceeds without all components in place, tracking of the remaining components is important to make sure that they are eventually delivered.

Once the life-cycle management organization has accepted delivery of the project, the project can now be closed.

Lessons Learned and Closeout

A lessons-learned session is a way to synthesize and assemble the knowledge that the project team has gained through a collective activity. In this session, the feedback collected is related to the team experience with the processes, tools, techniques, and project organization throughout the project. The objective is to code the experience gained in the form of recommendations that can be useful for future endeavors and to establish a "learning organization." The success of such a session depends on an atmosphere of trust devoid of blame and finger pointing. This is why such sessions are inappropriate for handling cases of incompetence or dereliction of duty. In that vein, the project manager can take advantage of anonymous feedback to start a meaningful exchange, provided that that anonymous information is not used to discuss personal flaws or to settle scores.

It is usually useful to have multiple lessons-learned sessions within each of the functional organizations that have participated in the project and at major milestones. Gathering the collective experience at the end of each subphase has several advantages. Memories are fresh and team members have not been engaged in other activities; in addition, the recommendations may be put to use in the remaining parts of the project. In addition, at the end to the project, there may not be the opportunity to capture all the lessons for the start. Thus, the final session is better reserved for the overall management of the project by making a focusing on the key points that have consistently been raised in the earlier sessions.

For lesson-learned sessions to be useful, the participation of as many members as possible should be encouraged. Therefore, it is better to include the session as an agenda item of a regular project meeting and remind team members to prepare their evaluations and suggestions. Starting with objective measurements of the project performance, the focus should be on specific successes, failures, and roadblocks, with the aim of modifying the processes to take advantage of the positive points and avoid the negatives.

The outcome of such a session would be a list of root causes of the problems or successes that have been experienced and a plan on how to deal with them.

Quality-of-Service Metrics

Service quality management depends on the quality of the user interface and the availability of meaningful service instructions and the ease of use. The various aspects related to management of the quality of service are shown in Figure 11.6.

The quality-of-service parameters cover the following aspects:

- The integrity of the service from both objective and subjective viewpoints. The actual parameters depend on the nature of the content (voice, video, transactional data, interactive data, etc.) and its tolerance to loss and delay.
- The user interface and the availability of meaningful service instructions and the ease of use. The metrics used are the frequencies of users making mistakes, the probabilities of giving up and abandoning their attempts to access the service, or the

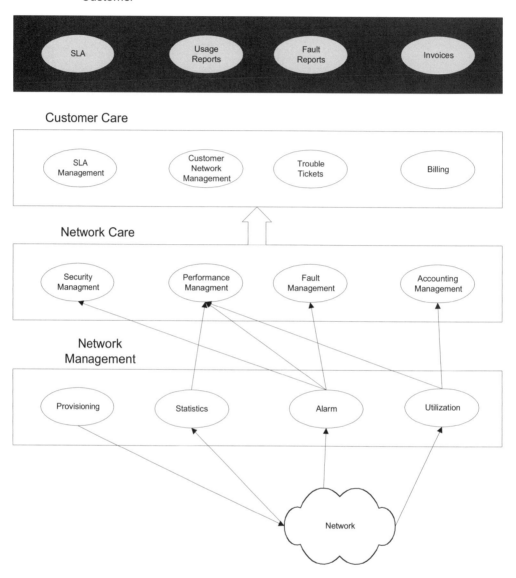

Figure 11.6 Organization of the quality of service elements.

probabilities of successful service completion. This factor covers the responsiveness to customer's requests or complaints in terms of the time to obtain a satisfactory answer, the percentage of requests answered, and the percentage of satisfied customers.

- Network-oriented parameters, which have traditionally called the focus of data networks provider.
- The operations management.

Customer Care Performance. Customer care is evaluated in terms of the time to obtain a satisfactory answer to a request or a complaint, the percentage of requests answered, and the percentage of satisfied customers. This aspect covers the service support functions of provisioning and maintenance as well as the charging and billing—for example, the mean service provisioning time, the mean time to repair, the transparency of the tariff structure, and the accuracy of billing.

Network Performance. The network-oriented performance parameters represent the combined contribution of the switching and transmission equipment. Depending of the type of the network, the quality of network service is measured in terms of availability, reliability, impacted user minutes (IUM), defects per million (DPM), and percent on-time order completion rate. The measures must relate directly to the user experience. These types of parameters are usually specified in service level agreements (SLA).

Based on measured network performance, trunk utilization analysis, switch engineering guidelines, and so on, it may need to reevaluate the network engineering rules. In particular, root cause analysis of outages can also force some changes, either in the operating procedures, in the engineering rules, or in the equipment qualification testing procedures.

OA&M Quality. This OA&M quality relates to the service features penetration, availability, and reliability of the overall service. Measures of such quality include the provisioning interval (the number of days after an order is made that the service is available), billing accuracy, the transparency of the tariff structure, the geographic availability, the reliability of the overall service, the mean time to repair, and so on.

BUSINESS AND NETWORK EVOLUTION

To maintain a sustainable competitive advantage, existing services must be enhanced by expanding their span, introducing new services, and/or evolving the network technologies. As the product portfolio expands, the product mix should be evaluated to produce a coherent evolution policy in synchrony with the sponsor needs, the market requirements, and technology evolution as described in Chapter 2. This entails adopting an integrated view bringing together technological inputs, customer request, marketing input, and the strategic plan (see Figure 11.7).

New product development involves market research and customer feedback in addition to technology-scanning. Their relative weights and the primary drivers depend on the type of innovation. This is described in terms of service and markets in the diagram of Figure 11.8. In this representation, the functional organizations are arranged in each quadrangle in the order of precedence in defining the service. For example, customer input is more important in the case of existing services (incremental innovations), while technology forecast is more important for new services using a new generation of technology (platform innovations). Likewise, market research is the leader in exploring new markets for existing services, which is also the case for architectural innovations; what marketers call "mid-life kickers" that can be used to reposition existing products into new markets. For radical innovations where new services and new markets are being created at once, a combination of technical and marketing associations through al-

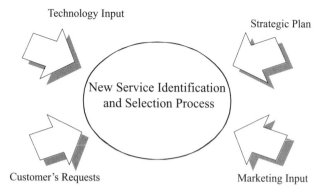

Figure 11.7 New service identification and selection process.

liances, consortia, and joint ventures may supply needed skill or help to explore new product ideas.

As was explained throughout the book, the technology S-curve provides considerable assistance in positioning the services on the technology life cycle and in anticipating technology shifts.

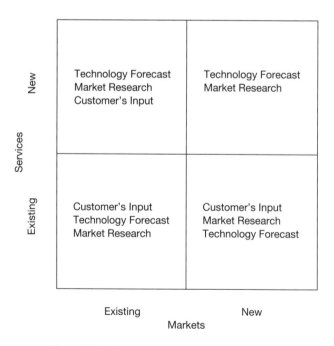

Figure 11.8 Market and service evolution matrix.

SUMMARY

Telecommunication is an exceptional member of the industrial family of information processing. It is built on the conjunction of machines, people, and procedures to deliver services on a continuous basis with a constant level of performance as per the service level contract. Parts of the business are technology-based; others are labor-intensive; however, the cost of operations and maintenance of the service dwarfs the expenses in equipment procurement or in real estate. Routine projects in telecommunication services aim at reducing the cost of operation and maintenance to increase return on investment. Project realization within the constraints of budget or schedule should not be at the expense of the service success in terms of ease of operation or the total cost over the lifetime of the service.

This chapter covered the relatively rare case of establishing a new network to offer a new service. Most projects in telecommunication services are incremental innovations to improve the day-to-day operations or to enhance the quality of existing services; that is, they constitute a subset of what was discussed in this chapter. Whether the overall project qualifies as a platform or a radical innovation depends on the degree of changes in any or all of the ingredients: networking technology, OSS and operating procedures, training, and human resource management. For example, if an enterprise decides to re-engineer its private network by consolidating the physical infrastructure and migrating to a new generation of technology to improve the performance of its business support processes, this would be a platform innovation. However, if it also decides to establish a competitive business and offer the new resources to outside parties, perhaps as a way of generating additional revenues, this would be a radical innovation: The management processes for an enterprise network are totally different from those that would be used to support paying customers. At a minimum, adequate billing and customer care functions would need to be added. As we have shown in the previous chapters, the correct classification of the innovation category is an academic exercise; it has profound consequences on the type of organizations that should participate in defining the scope of the project, the nature of project organization, and the approach for managing the various aspects of the project (risks, vendors, quality, communication, etc.).

APPENDIX

The defects per million (DPM) is a measure for service availability in packet networks that is compatible with traditional measures in circuit-switched networks. For frame relay and asynchronous transfer mode (ATM) networks, the DPM for permanent virtual circuits (PVCs) is defined as

$$\text{DPM} = \frac{\text{Sum of failed PVCs} \times \text{duration of failures in hours}}{\text{Sum of PVCs available in the network to date} \times \text{hours in the year to date}} \times 10^6,$$

while for IP networks it is defined as

$$\text{DPM} = \frac{\text{Sum of ports experiencing a failure} \times \text{duration of failures in hours}}{\text{Sum of active ports available in the network to date} \times \text{hours in the year to date}}$$
$$\times 10^6.$$

12

SOME FINAL THOUGHTS

This chapter presents some reflections on the following aspects of telecommunication services:

- Continuity and change
- Project success and service success
- Competition and government policies
- Standardization
- Outsourcing

CONTINUITY AND CHANGE

Telecommunication services are built on technologies and systems to serve subscribers in their social settings. Management of these interactions depend on networking technologies, operation support systems (OSS), and operational procedures. Accordingly, the economies of scale in telecommunication services depend equally on the volume of traffic and on the operational efficiencies gained through automation and standardization.

Service management and network management in telecommunications involve highly specialized knowledge that can be gained only by capitalizing on past experience. The systematic codification of such knowledge is essential to increase efficiency and to optimize expenditures. Potential entrants cannot readily replicate such expertise, no matter what their size and their financial backing. Not only do they face the typical issues of brand identification and customer attraction, they also have to develop formal processes for operating their services, managing their suppliers, assisting their customers, and building their expertise from the bottom up; that is, they have inherently higher costs for the same service quality than established firms.

Managing Projects in Telecommunication Services. By Mostafa Hashem Sherif
Copyright © 2006 The Institute of Electrical and Electronics Engineers, Inc. **223**

As long as the environment is stable, established operators have many advantages over potential entrants, unless these industry leaders are mismanaged or there is a political will to favor new entrants at any cost. Most of the innovations in telecommunication services take place within the framework of a dominant design, through the accumulation of incremental steps to improve the service offers in terms of affordability, ease of use, availability, and so on. When new needs arise that cannot be satisfied with the existing order but the technologies needed are available, a relatively short burst of "noncumulative development episodes" [Kuhn, 1970, p. 92] could lead to a new stable point. The formulation of Abernathy and Clark [1985] provides a way to understand how the successive equilibria are reached. By considering two factors—the technology and the value chain—we can identify four main categories of innovations. Each category has its own set of requirements in terms of capital, technical know-how, infrastructure development, and organizational design. In the case of high-end services using new platforms, the aspiring operators must spend heavily to retrain customers, manage their services and/or their networks, test ancillary equipment, qualify their suppliers, and build their expertise with all the aspects of the service over a long period of time. Architectural innovations may offer an entry for services that are not mission critical, provided that the incumbent service providers do not lower their prices below the level at which their competitors can be profitable. These observations provide some explanation as to why very few alternative operators have successfully displaced the incumbent telephone companies even when legislation has forced the incumbents to provide local access to their competitors.

When high-end and low-end services compete based on different technologies, such as circuit-switched voice and voice over IP, the end-user evaluation criteria may evolve if new applications can complement the lower-quality services. However, the chances to leapfrog the industry leader are larger in Greenfield areas—that is, where a new service is created *ex nihilo* because radical innovations level the playing field by making most prior expertise irrelevant. In such a major shift, past solutions to optimize the operations and the network management often end up by making the existing infrastructure less flexible and less adaptable. As a consequence, such shifts could reverse the balance of power and destabilize established firms.

Even though it is tempting to advise companies to disrupt the market place through radical innovations [Atkinson, 2003], there are limitations as to what a single entity can do. Radical innovations in telecommunication services depend on the modification of the whole milieu, a process that takes at least 15 years to mature. In such a case, defining the customer requirement upfront is very difficult except in limited cases, such as for government contracts.

The implication of all of these considerations is that routine service projects in telecommunications are usually the most profitable and have the highest chances of success. Explorative projects can still be successful if the stakeholders have a good grasp of the environment in which they operate.

PROJECT SUCCESS OR SERVICE SUCCESS?

The criteria for evaluation project success typically rely on the triad of scope, schedule, and cost. Yet because the product of a project in telecommunication services is the service itself, deficiencies in the project implementation may plant the seed of the inability of the service to meet expectations. The pressure for meeting the schedule or reducing cost may lead to short cuts with catastrophic consequences, because a significant portion of a service comes from operational considerations. The failures of some of the satellite-based wireless communication networks (e.g., iridium) originated from poor demand forecast due to insufficient research or misinterpretation of customer preferences and the inability to meet technical specifications [Ahn et al., 2005]. Table 12.1 summarizes the main contributors of project failures in telecommunication services per innovation type.

COMPETITION AND GOVERNMENT POLICIES

In his classical book on competitive strategy, Michael Porter considered five factors: the power of suppliers and customers, the threats of new entrants and substitutes and the rivalry among existing competitors; government action was not seen as a force in itself. In this model, governments affect the competitive landscape as buyers, or suppliers or arbitrators that set the limits on the competition, through financial and environmental regulations, tax incentives, research grants as well security laws [1980, p. 29]. Such a description, however, does not take proper account of the essential role that governments play in telecommunication services.

The ubiquity of telecommunication services (starting from telegraphy and then telephony) did not happen by accident. (We should always keep in mind that the U.S. Federal Government subsidized the development of the Internet for more than 15 years.) Governments have been actively promoting telecommunication services because, through network externalities, their value increases with the number of subscribers and because their effects are beneficial to all sectors of the society. In addition, the telecommunications infrastructure would have to be operational in the case of national emergencies or for military purposes. In general, government policies have to tended to address the following items [Marcus, 2005]:

- The basic obligations that service providers have to meet
- The types of services that should be subsidized and the level of subsidies
- The source of the subsidies and the way they are distributed

Table 12.1 Main Causes of Project Failure in Telecommunication Services

Innovation Type	Causes of Failure
Radical	Lack of expertise (technical and operational)
	Technology does not meet service requirements
	Poor demand forecast
	Lack of sponsor or government support
Platform	Failure to scale (technological limitation, lack of service delivery capability, organizational limitations, etc.)
	Poor demand forecast
Architectural	Ineffective marketing
	Lack of customer's support
Incremental	Misinterpretation of customer preferences
	Deficiencies in project implementation

In the last two decades, competition was added to the above list. The deregulation of telecommunication services started first in the United Kingdom in 1981 and then proceeded in earnest with the divestiture of the Bell System in the United States in 1984. In 1985, the White Book of the European Economic Community (EEC) pushed for the deregulation of the telephone market in Europe to be fully implemented by 1998. Of course, this wave did not happen by magic. The liberalization of the U.S. telecommunication market created a significant balance-of-trade deficit in both equipment and services. Foreign manufacturers could now sell their equipment to U.S. operators even though the procurement processes of national telephone monopolies erected non-tariff trade barriers against U.S. manufacturers. Meanwhile, as competition among U.S. operators drove down the price of outbound international calls from the United States but not in the reverse direction, call-back operators stepped in to connect parties outside the United States and make it appear as if their calls had originated from the United States. While they offered end-users considered cost savings, these new architectural services increased the unbalance of minutes of traffic from and into the United States. The consequence was that, according to the rules for accounting and settlement of international calls, U.S. operators had to disburse more than what they received from their foreign correspondents. As a result, by the early 1990s, the U.S. annual balance-of-payments deficit on telecommunication services reached about $3 billion. At a time of concern over the growth of the federal budget deficit, the Federal Communications Commission (FCC), the State Department, the Commerce Department, and Congress put considerable pressure to revamp the rules of international trade in telecommunication services. To redress this situation, U.S. trade representatives pushed for the breakup of state monopolies to open foreign markets to U.S. telecom manufacturers and to bring down the price of international services through competition [Cowhey and Aronson, 1993, pp. 185–196]. As the deregulation wave spread to most countries, vertically integrated service providers were replaced with open interfaces and more transparent cost-based pricing. A negative consequence of the new competitive environment has been the generation of large mismatches between the benefits to the economy as a whole and the revenues and expenditures of telecommunication operators in addition to geographical and social unbalances. Eventually, some political intervention will be needed to redress these unbalances.

Government policies still affect telecommunication services directly and indirectly through taxation, by the licensing process for operation and interconnection, or through the legislative process. The process of frequency allocation gives authorities absolute control over the number and the timing of entries in wireless and cellular communications. In the United States, established television broadcasters delayed the re-allocation of underutilized frequencies to mobile communications from 1946 to 1975; in the meantime, cellular services were readily adopted in Europe and Japan [West, 2000]. Another example is the CI-2 service for cordless access and wireless PBX that was designed in the United Kingdom and then adopted by the European Telecommunication Standards Institute (ETSI) [Pandya, 2000, p. 90]. While the Korean government approved a CI-2 trial license in December 1993, it withheld commercial approval until June 1996, at which time the service had lost its value compared to cellular services [Ahn et al., 2005]. To illustrate the indirect consequences of government action, consider the case of digital spectrum auctions, which started in the United States in 1994 and then reached a record amount when the United Kingdom auctioned five licenses for third-generation (3G) wireless spectrum raising $35 billion in April 2000. These amounts whetted the appetites of other countries so that by the end of that year a half-dozen such auctions were conducted for the combined amount of $100 billion leaving at the end the European wireless operators with drained coffers and large loans [Anderson, 2002]. This stirred debates on the justification of auctions for spectrum allocation, how to run them, and whether IEEE 802.11b (Wi-Fi) services that use the unregulated spectrum usher a new way where "technology helps the markets self regulate."

Government policies can also favor one technology over the other. While voice services on the Public Switched Telephone Network (PSTN) are highly regulated and subject to taxes, voice over IP services are not. Government intervention is also evident through the International Telecommunication Union (ITU). Its radio sector defines the rules of frequency allocation through international treaties among governments. In the ITU telecommunication standardization sector, private firms define the technical rules within the parameters that governments agreed to for international telecommunications.

Content control, rules on encryption, and censorship affect the commercial viability of services. Intellectual property rights (IPR), patent laws, and copyright laws such as the treaties of the World Intellectual Property Organization (WIPO) put limits on the ability to copy digital productions and prohibit any action to intentionally circumvent the technical means to protect copyrights and related rights.

Another aspect that needs to be revisited in Porter's model relates to the relationship among competing telecommunication service providers. Their rivalry does not prevent them from cooperating by exchanging traffic on reasonable terms and conditions to take advantage of the network externalities (i.e., the value of the service offered increases with the member of subscribers). It is extremely rare when competing telephony networks refuse to exchange traffic under reasonable terms. This is more common in the case of Internet service providers that do not properly maintain their networks to prevent spam or to apply the necessary patches to protect from security attacks.

STANDARDIZATION

Because the telecommunication industry is dynamic and constantly evolving, networks are inevitably heterogeneous. Without interoperability and interface standards, it would have been impossible to offer transparent telecommunication services end-to-end across different networks and diverse equipment. Many innovations could not have worked without standards, such as national and international roaming—that is, the use of the mobile terminal outside the home base station or the home region of the network operator. Of course, operators may opt to enrich their service with nonstandard features, but this may come at the cost of smooth evolution. A case in point is that of the U.S. wireless operator Nextel, which attracted many small business owners and utility workers with its "Push-to-talk" walkie-talkie service for direct connection using a proprietary technology jointly developed with Motorola. Unfortunately, this also injected considerable difficulties in the migration toward the new technology for 3G services.

The production of technical specification intertwines technical considerations with business, social, political and ideological factors. This is especially true in the cases of information and telecommunication technologies (ICT) which are embedded in a wide variety of products and services. ICT standards attempt to meet a multiplicity of needs so that there is an inevitable tension among the different requirements, particularly because of the differences in the time horizons of manufacturers and service providers [McCalla and Whitt, 2002; Sherif, 2003b]. Successful standardization requires the identification and the articulation of the most important issues to be addressed. In the case of incremental and platform innovations, it is relatively easier to anticipate the expectations of the users, whether the subscribers or the on-site workforce of the network operators or service providers. This is much more challenging, however, in the case of radical innovations, such as GSM, given that the average time for the corresponding standard to be fully and successfully implemented is around 8–10 years [Haug, 2002]. One possible approach to ensure the constancy of purpose is to maintain technological neutrality and avoid being tied to one specific implementation—that is, to consult with the manufacturers to understand where the technology is going but to protect the standardization schedule and agenda from their influence. By relating the type of standards to the technology life cycle, it may also be possible to define the scope of standardization and the right amount of cooperation with manufacturers.

OUTSOURCING

New communication and information technologies have allowed companies to split up tasks and to disperse the locations of their execution. First, manufacturers delocalized their factories or sourced their components from a variety of global producers. Likewise, service industries are currently splitting functions, such as accounting or customer relations, and distributing them to providers on a worldwide basis. This distribution ensures round-the-clock operation and increases the performance by seeking expertise wherever it may be. Outsourcing has been used to off-load fringe activities to specialized providers, to reduce labor costs, or as an anti-union strategy. Today, it is estimated that around 75% of large companies throughout North America and Europe have outsourced the management of human resources.

One effect of outsourcing is to increase the need for reliable and affordable telecommunication services. However, telecommunications companies have also outsourced their internal operations such as payroll and pension plan administration, human resources administration, and information technology (IT) support. One possible way to figure out the limit of outsourcing is Porter's analysis of the relative power of suppliers in a competitive environment. Such an analysis indicates that suppliers exercise considerable influence if [Porter, 1980, pp. 27–28]

1. Their number is limited.
2. The cost of switching from one supplier to another is substantial.
3. Their contribution is essential to the buyer's business, while the latter has limited control over the parameters of the delivery (short of litigation).

With these factors in mind, outsourcing key parts of the operations such as field support is a strategic step that should not be undertaken lightly in the hope of a quick fix and without consideration of all the tangible and intangible costs associated with the outsourcing decision. In the long-term, outsourcing could prevent the accumulation of in-house expertise on the limitations of the current technology and processes and hence the ability to forecast technology or markets and prepare for the future.

REFERENCES

W. J. Abernathy and V. B. Clark, Mapping the winds of creative destruction, *Research Policy* **14**(1):2–22, 1985.

AFITEP, *Le Management de Projet—Principes et Pratique,* 2nd edition, AFNOR, Paris, 2000.

J.-H. Ahn, M.-S. Kim, and D.-J. Lee, Learning from the failure: Experiences in the Korean telecommunication market, *Technovation* **25**(1):69–82, 2005.

H. Akhavan, Why synchronizing systems is a tough call, *Financial Times FT-IT Review,* March 31, 2004, p. 2.

S. R. Ali, Digital switching system testing—Some special considerations, *IEEE Global Telecommunications Conference, Globecom'95* **2**:1221–1224, 1995.

C. Anderson, Winner's curse: The 3G auctions were the last party of an old regime, *Wired* **10**(5):61, 2002.

ANSI/EIA-748-1998, *Earned Value Management Systems,* Electronics Industry Alliance, Arlington, VA, June 1998.

ANSI/ISO/ASQC Q9001-1994, Quality Systems—Model for Quality Assurance in Design, Development, Production, Installation, and Servicing, American Society for Quality, 1994.

G. R. Ash and P. Chemouil, 20 years of dynamic routing in circuit-switched telephone networks: Looking backward to the future, *IEEE Communications Magazine, Global Communication Newsletter,* October 2004. Also http://www.comsoc.org/dl/gcn/gcn1004.html.

G. Ash and G. Choudhury, PNNI routing congestion control, *IEEE Communications Magazine* **42**(11):154–160, 2004.

G. Asselin and R. Mastron, *Au Contraire: Figuring Out the French,* Intercultural Press, Yarmouth, ME, 2001.

W. Atkinson, Beyond the basics, *PM Network,* **17**(5):38–43, 2003.

P. Awde, Operators late to catch the wave, *Financial Times, FT-IT Review,* December 1, 2004, p. 3.

B. Barthélemy, *Gestion des Risques: Méthode d'Optimisation Globale,* Éditions d'Organisation,

Paris, 2000.

C. Bergren, J. Söderlund, and C. Anderson, Clients, contractors, and consultants: The consequences of organizational fragmentation in contemporary project environments, *Project Management Journal* **32**(3):39–48, 2001.

F. Betz, *Strategic Technology Management,* McGraw-Hill, New York, 1993.

R. Bhandari, *Survivable Networks—Algorithms for Diverse Routing,* Kluwer Academic Publishers, Boston, 1999.

BITS, BITS Guide to business-critical telecommunications services, Washington, D.C., November 2004. Available at www.bitsinfo.org.

K. Black, Causes of project failure: A survey of professional engineers, *PM Network* **10**(10):21–24, 1996.

J.-P. Breuer and P. de Bartha, La médiation interculturelle au secours des partenaires franco-allemands, *Gérer et Comprendre, Annales des Mines* **30**:50–58, 1993.

C. Brown-Humes, Ericsson to run Hutchinson arm, *Financial Times,* December 3, 2002, p. 22.

C/SSR, *Cost/Schedule Status Report (C/CCSR) Joint Guide,* Departments of the Army, the Navy and the Air Force; the Ballistic Missile Defense Organization, the Defense Logistics Agency; the Defense Contract Audit Agency; and the National Security Agency, May 1, 1996.

S. Calé, *La Gestion des Projets Télécoms,* Lavoisier, Paris, 2005.

A. Cardon, *Jeux de Manipulation—Petit Traité des Stratégies d'Échec qui Paralysent nos Organizations,* Les Éditions d'Organisation, Paris, 1995.

E. Carmel, *Global Software Teams: Collaborating Across Boundaries and Time Zones,* Prentice-Hall, Englewood Cliffs, NJ, 1999.

D. Cavendish, H. Ohta, and H. Rakotoranto, Operation, Administration, and Maintenance in MPLS networks, *IEEE Communications. Magazine* **42**(10):91–99, 2004.

CCITT, *Handbook on Quality of Service and Network Performance,* Geneva, 1993.

S. Chevrier, Le management de projets interculturels: Entre le rêve du melting pot et le cauchemar de la tour de Babel, *Gérer et Comprendre, Annales des Mines* **45**:38–47, 1996.

C. M. Christensen, *The Innovator's Dilemma: When New Technologies Cause Great Firms to Fail,* Harvard Business School Press, Boston, 1997.

T. Cleary, Translator's introduction to Sun Tzu's *The Art of War,* Shambala, Boston, 1988.

P. F. Cowhey and J. D. Aronson, *Managing the World Economy, The Consequences of Corporate Alliances,* Council on Foreign Relations Press, New York, 1993.

P. B. Crosby, *Quality is Free. The Art of Making Quality Certain,* McGraw-Hill, New York, 1979.

P. Curwen, Survival of the fittest: Formation and development of international alliances in telecommunications, *Info* 1(2):141–158, 1999.

M. Daoud, *The Processing of EST Discourse: Arabic and French Native Speaker's Recognition of Rhetorical Relationships in Engineering Texts,* Ph.D. Dissertation, University of California, Los Angeles, 1991.

T. De Marco and T. Lister, *Peopleware: Productive Projects and Teams,* Dorset House, New York, 1987.

W. E. Deming, *Out of the Crisis,* Massachusetts Institute of Technology, Center for Advanced Engineering Study, Cambridge, MA, 1986.

C. L. Desmond, *Project Management for Telecommunications Managers,* Kluwer Academic Publishers, Norwell, MA, 2004.

J. Dix, AT&T to cut Net 1000? *Communication World,* February 20, 1985.

E. Dunn, Does ATM need MPLS? Vertical Systems Group, ATM Forum Presentation, January 21, 2002, San Diego, CA.

S. C. Dunn, Motivation by project and functional managers in matrix organizations, *Engineering Management Journal* **13**(2):3–9, 2001.

T. Egyedi and A. G. A. J. Loeffen, Succession in standardization: Grafting XML onto SGML, *Proceedings of 2nd IEEE Conference on Standardization and Innovation in Information Technology* (SIIT 2001) **Oct. 3–6**:38–49, 2001.

L. Fang, A. Atlas, F. Chiussi, K. Kompella, and G. Swallow, LDP failure detection and recovery, *IEEE Communications Magazine* **42**(10):117–123, 2004.

Q. W. Fleming and J. M. Koppelman, *Earned Value Project Management*, 2nd edition, Project Management Institute, Newton Square, PA, 2000.

Forrester Research, *European Incumbents Telco's VoIP Road Map*, October 2003, www.forrester.com.

J. D. Frame, Requirements management: Addressing customer needs and avoiding scope creep, in *Project Management for Business Professionals: A Comprehensive Guide*, J. Knutson, ed., John Wiley & Sons, New York, 2001, pp. 63–80.

Futuribles, Disparités des pratiques d'information en Europe, **45**:69, 1999.

J. Gapper, The advantage of ruling the world, *Financial Times* December 23, 2004, p. 11.

J. Garcia-Arreola, *Technology Effectiveness Audit Model: A Framework for Technology Auditing*, Master's thesis, University of Miami, 1996.

T. Gezo, M. Oliverson, and M. Zick, Managing global projects with virtual teams, *Proceedings of the 30th Annual Project Management Institute Symposium*, Philadelphia, 1999.

M. Goddard and G. Klein, Correspondence, *Project Management Journal* **32**(3):62, 2001.

S. Golinski and P. J. Rutkowski, Flawless execution: The 1B processor project: 1996 Project of the Year, *PM Network* **11**(2):19–29, 1997.

J. C. Goodpasture, *Managing Projects for Value*, Management Concepts, Vienna, VA, 2002.

L. Gorchels, Transitioning from engineering to product management, *Engineering Management Journal* **15**(4):40–49, 2003.

E. T. Hall, *The Silent Language*, Anchor Books/Doubleday, New York, 1959.

E. T. Hall, *Beyond Culture*, Anchor Books/Doubleday, New York, 1976.

E. T. Hall and M. R. Hall, *Understanding Cultural Differences*, Intercultural Press, Yarmouth, ME, 1990.

M. Hansen, Turning the lone star into a real team player, *Financial Times, FT Summer School, Day 4: Knowledge Management* August 8, 2002, p. 8.

T. Haug, A commentary on standardization practices: Lessons from the NTM and GSM mobile telephone standards history, *Telecommunications Policy* **26**(3–4):101–107, 2002.

G. T. Haugan, *Effective Work Breakdown Structures*, Management Concepts, Vienna, VA, 2002.

J. Hauschildt, G. Keim, J. W. Medcof, Realistic criteria for project manager selection and development, *Project Management Journal* **31**(3):23–22, 2000.

R. M. Henderson and K. B. Clark, Architectural innovation: The reconfiguration of existing project technologies and the failure of established firms, *Administrative Science Quarterly* **35**:9–30, 1990.

D. A. Hoeflin, A Second Look at When Should One Stop Product Testing, AT&T Memorandum, 2000.

D. Hoeflin and M. H. Sherif, An integrated deficit tracking model for product deployment in telecom services, in *Proceedings of the 10th IEEE Symposium on Computer and Communications* ISCC 2005, 27–30 June 2005, Cartagena, Murcia, Spain, pp. 927–932.

G. Hofstede, *Culture's Consequences: International Differences in Work-Related Values*, 1980, abridged edition, Sage Publications, 1990.

G. Hofstede, *Culture's and Organizations: Software of the Mind—Intercultural Cooperation and Its Importance for Survival,* McGraw-Hill, New York, 1997.

P. Holahan and A. C. Mooney, Conflict in project teams: Gaining the benefits, avoiding the costs, *Current issues in Technology Management, Stevens Alliance for Technology Management* **8**(3):00–00, 2004.

K. Hope, Motorola picks up Athens telecoms baton, *Financial Times* December 10, 2003, p. 6.

C. L. Howe, Another tangled network, *Datamation* **36**(6):64–73, 1986.

J. M. Howell, Organization contexts, charismatic and exchange leadership, in Kellogg Leadership Studies Project *KLSP: Transformational Leadership Working Papers,* Academy of Leadership Press, University of Maryland, College Park, MD, 1997, http://www.academy.emd.edu/publications/klspdocs/howell.html.

IEEE Std 1044.1-1995, *IEEE Standard Classification for Software Anomalies,* Software Engineering Standards Committee of the IEEE Computer Society, New York.

P. d'Iribarne *La logique de l'Honneur: Gestion des Enterprises et Traditions Nationales,* Seuil, Paris, 1989.

P. d'Iribarne, Coopérer à la belge: La mise en œuvre problématique d'un agenda électronique, in *Cultures et Mondialisation: Gérer par-delà les Frontières,* Seuil, Paris, 1998, pp. 41–62.

International Telecommunication Union—Recommendation E.800, Terms and definitions related to quality of service and network performance including dependability, August 1994.

International Telecommunication Union—Recommendation M.495, Use of circuits for voice-frequency telegraphy, November 1988.

International Telecommunication Union—Recommendation M.3208.1, TMN management services for dedicated and reconfigurable circuits network: Leased circuit services, October 1997.

International Telecommunication Union—Recommendation X.137, Availability performance values for public data networks when providing international packet-switched services, August 1997.

International Telecommunication Union—Recommendation X.140, General quality of service parameters for communication via public data networks, September 1992.

International Telecommunication Union—Recommendation X.160, Architecture for customer network management service for public data networks, October 1996.

International Telecommunication Union—Recommendation X.161, Definition of customer network management services for public data networks, August 1997.

International Telecommunication Union—Recommendation X.162, Definition of management information for customer network management service for public data networks to be used with the CNMc interface, March 2000.

International Telecommunication Union—Recommendation X.800, Security architecture for Open Systems Interconnection for CCITT applications, March 1991

International Telecommunication Union—Recommendation X.790, Trouble management function for ITU-T applications, November 1995.

ISO/IEC-12207 Information Technology—Software cycle process, 1995.

M. Jablonowski, Prioritizing disaster recovery plans using risk maps, *Disaster Recovery Journal* **13**(3):58–60, 2000.

K. A. Jehn, Diversity, conflict and team performance: Summary of program research, *Performance Improvement Quarterly* **12**(1):7–19, 1999.

K. A. Jehn, G. B. Northcraft, and M. A. Neale, Why differences make a difference: A field study of diversity, conflict and performance in workgroups, *Administrative Science Quarterly* **44**:741–763, 1999.

Z. Jelinski and P. Moranda, Software reliability research, in *Statistical Computer Performance Evaluation,* 1972, W. Freiberger, ed., Academic Press, New York, 1972, pp. 465–484.

N. O. Johannesson, The ETSI computation model. A tool for transmission planning of telephone networks, *IEEE Communications Magazine* **35**(1): 70–79, 1997.

D. Jin, *The Dynamics of Knowledge Regimes: Technology, Culture and National Competitiveness in the USA and Japan,* Continuum, London, 2001.

D. Jolly, The Vizzavi joint venture: An exogamic marriage between Vivendi and Vodafore as a way to entering the e-economy, in *Management of Technology, Internet Economy: Opportunities and Challenges for Developed and Developing Regions of the World, Selected Papers from The 11th International Conference on Management of Technology,* Y. A. Hosni and T. M. Khalil, eds, Elsevier, Oxford, 2004, pp. 55–67.

C. Jones, *Assessment and Control of Software Risks,* Yourdon Press, PTR Prentice-Hall, Englewood Cliffs, NJ, 1994.

J. Kay, A failing brand lives on borrowed time, *Financial Times* August 17, 2004, p. 13.

T. Keil, De-facto standardization through alliances—lessons from Bluetooth, *Telecommunications Policy* **26:**205–213, 2002.

H. Kerzner, *Project Management: A Systems Approach to Planning, Scheduling and Controlling,* 6th edition, John Wiley & Sons, New York, 1998.

T. Khalil, *Management of Technology: The Key to Competitiveness and Wealth Creation,* McGraw-Hill, Boston, 2000.

T. J. Kloppenborg and J. A. Petrick, *Managing Project Quality,* Management Concepts, Vienna, VA, 2002.

J. M. Koppelman and Q. W. Fleming, Earned value management: An introduction, in *Project Management for Business Professional—A Comprehensive Guide,* J. Knutson, ed., John Wiley & Sons, New York, 2001, pp. 166–178.

J. Knutson, ed., *Project Management for Business Professional—A Comprehensive Guide,* John Wiley & Sons, New York, 2001.

J. Knutson and I. Bitz, *Project Management: How to Plan and Manage Successful Projects,* American Management Association, New York, 1991.

O. Kroeger and J. M. Thuesen, *Type Talk—or How to Determine Your Personality Type and Change Your Life,* Delacorte Press, New York, 1988.

D. W. Krumwiede and J. P. Lavelle, The effect of top-manager personality on a total quality management environment, *Engineering Management Journal* **12**(2):9–14, 2000.

T. S. Kuhn, *The Structure of Scientific Revolutions,* 2nd edition, The University of Chicago Press, Chicago, 1970. A very good outline and study guide of this book by Professor Frank Pajares of Emory University is available at www.emory.edu/EDUCATION/mfp/Kuhn.html.

J. Lapierre and B Hénault, Bidirectional information transfer: An imperative for network and marketing integration in a Canadian telecommunications firm, *Journal of Product and Innovation Management* **13:**152–166, 1996.

R. Larini, A new company for state E-ZPass, *Star Ledger,* October 20, 2001, p. 11.

R. Le Maistre, Another French revolution? *Lighreading.com,* March 31, 2004, http://www.lightreading.com/document.asp?site=lightreading&doc_id=50319.

D. R. Lee, G. A. Bohlen, and P. J. Sweeney, A comparison of U.S. and European project manager decision-making styles, *Engineering Management Journal* **7**(3):25–32, 1995.

S. Liebesman, A. Jarvis and A. V. Dandeka, *TL 9000—A Guide to Measuring Excellence in Telecommunications,* ASQ Quality Press, Milwaukee, 2001.

D. Limerick and B. Cunnington, *Managing the New Organization: A Blueprint for Networks and Strategic Alliances,* Jossey-Bass Publishers, San Francisco, 1993.

M. R. Lyu, ed., *Handbook of Software Reliability Engineering,* McGraw-Hill, New York, 1996.

J. Malinconico, E-ZPass grows despite its $1 monthly fee, *Start Ledger,* October 3, 2002, http://www.nj.gov/transportation/newsletter/2002/oct/ezpass.pd.

E. Mankin, Why bad projects are so hard to kill, Feb. 10, 2003, http://mysite.verizon.net/vze4dvjj/id11.html.

J. S. Marcus, Universal service in a changing world, *IEEE Communications Magazine* **43**(1):16–17, 2005.

D. Martin, *AT&T and the Hard Lessons Learned from the Telecom* Wars, Anacom, New York, 2005.

A. Maslow, *Motivation and Personality,* 2nd edition, Harper & Row, New York, 1970. A good summary by Robert Gwynne is available at http://web.utk.edu/~gwynne/maslow.HTM.

J. S. Mayo, A new paradigm for R&D, presentation at the SATM Education Working Council, April 13, 1994, Murray Hill, NJ.

C. McCalla and W. Whitt, A time-dependent queueing network model to describe the life-cycle dynamics of private-line telecommunication services, *Telecommunications Systems* **19**(1):9–38, 2002.

D. Z. Milosevic, *Project Management Toolbox: Tools and Techniques for the Practicing Project Manager,* John Wiley & Sons, Hoboken, NJ, 2003.

W. R. Minor, Stranger in a stranger land: The American project manager working abroad, *PM Network* **13**(2):31–34, 1999.

G. Morgenson, Deals within telecom deals, *New York Times,* August 25, 2002, Sec. 3, pp. 1 and 10.

T. Moynihan, Coping with "Requirements-uncertainty": The theories-of-action of experienced IS project managers, in *Proceedings of the* International Conference of the Information Resource Management Association: "Challenges of Information Technology Management in the 21st Century," Anchorage, Alaska, May 21–24, 2000, M. Khosrowpour, ed., Idea Group, Hershey, PA, 2000, pp. 116–120.

J. Musa, *Software Reliability Engineering,* McGraw-Hill, New York, 1999.

J. Musa, A. Iannino, and K. Okumoto, *Software Reliability: Measurement, Prediction, Application,* McGraw-Hill, New York, 1987.

I. N. Myers, M. H. McCauley, N. L. Quenk, and A. L. Hammer, *MBTI® Manual. A Guide to the Development and Use of the Myers–Briggs Type Indicator,* 3rd edition, Consulting Psychologists Press, Palo Alto, CA, 1998.

G. Nairn, Operators make more out of less legacy system, *Financial Times, FT-IT Review,* April 14, 2004, 4.

M. Nakamoto, NTT DoCoMo, *Financial Times, World's Most Respected Companies,* December 17, 2001, p. IV.

M. A. Neale, G. B. Northcraft, and K. A. Jehn, Exploring Pandora's box: The impact of diversity and conflict on work group performance, *Performance Improvement Quarterly* **12**:(1):113–126, 1999.

New Jersey Department of Transportation, E-Z Pass: A plan that works for NJ—Honest and accountable, www.state.nj.us/transportation/ezpass/ezpass.pdf, July 11, 2002.

E. Noam, *Telecommunications in Europe,* Oxford University Press, New York, 1992.

A. Oodan, K. Ward, C. Savolaine and, P. Hoath, *Telecommunications Quality of Service Management: From Legacy to Emerging Services,* Institute of Electrical Engineers, Herts, United Kingdom, 2003.

W. J. Orlikowski, Learning from notes: Organizational issues in groupware implementation, in *Computerization and Controversy: Value Conflicts and Social Choices,* 2nd edition, Academic Press, San Diego, 1996, pp. 173–189 (reprinted from *The Information Society* **9**:237–250).

R. Orr, Move to reduce costs can also lose customers, *Financial Times, Special report on risk management,* June 2, 2004, p. 5.

R. Pandya, *Mobile and Personal Communication Systems and Services,* IEEE Press, Piscataway, NJ, 2000.

M. Pesola, Network protection is a key stroke, *Financial Times, FT Business Continuity,* March 9, 2004, p. 2.

PMI (Project Management Institute), *PMBOK® guide—A guide to the project management body of Knowledge,* Upper Darby, PA, 2000. This is also an ANSI standard ANSI/PMI 99-001-2000).

M. E. Porter, *Competitive strategy—Techniques for analyzing industries and competitors,* The Free Press, New York, 1980.

F. A. Prince, Why NASA's management does not believe the cost estimate, *Engineering Management Journal* **14**(1):7–12, 2003.

Probe Research, *Global Carriers: Competition, Networks and Markets,* **1**(1):January 2000.

P. F. Rad, *Project Estimating and Cost Management,* Management Concepts, Vienna, VA, 2002.

S. Ramasamy and E. Radwan, Tolls road fare collection systems, *12th International Conference on Management of Technology,* Paper 451, Track 8—Emerging and breakthrough technologies, May 13–15, 2003, Nancy, France.

T. Raz and S. Globerson, Effective sizing and content definition of work packages, *Project Management Journal* **29**(4):17–23, 1998.

R. F. Rey, ed., *Engineering and Operations in the Bell System,* 2nd edition, AT&T Bell Laboratories, 1983.

C. Ricci, AT&T plan to market a computer network hits snags repeatedly, *Wall Street Journal,* July 13, 1984.

S. R. Rich and D. E. Gumpert, *Business Plans that Win $$$—Lessons from the MIT Enterprise Forum^{SM},* Harper & Row, New York, 1985.

D. Richmar, The fall of AT&T Wireless, *Seattle Post-Intelligencer,* September 21, 2004, http://seattlepi.nwsource.com/business/191742_attw21.html.

C. Roche, Le management des connaissances dans des unités opérationnelles: L'exemple du service client par téléphone, *Réalités industrielles—Annales des Mines,* **Nov.**:87–91, 1998.

I. Royer (a), Why bad projects are so hard to kill, *Harvard Business Review,* **Feb.**:49–56, 2003.

I. Royer (b), When bad ideas won't die, *Harvard Business Working Knowledge,* March 24, 2003, http://hbswk.hbs.edu/pubitem.jhtml?id=3390&t=leadership.

P. S. Royer, Risk management: The undiscovered dimension of project management, *Project Management Journal* **31**(1):6–13, 2000.

M. F. Rubenstein, *Tools for Thinking and Problem Solving,* Prentice-Hall, Englewood Cliffs, NJ, 1986.

A. Sadeh, D. Dvir, and A. Shenhar, The role of contract type in the success of R&D defense projects under increasing uncertainty, *Project Management Journal* **31**(3):14–22, 2000.

M. H. Sherif, Diversity, culture and technical project management, *Proceedings of the 9th International Conference on Management of Technology,* February 21–25, Miami, Florida, 2000, TR 10.4A. Updated version is in *Management of Technology: The Key to Prosperity in the Third Millennium,* T. M. Khalil, L. A. Lefebvre, and R. M. Mason, eds., Elsevier, Oxford, 2001, pp. 373–379. Also at www.iamot.org/paperarchive/104a.pdf.

M. H. Sherif, A framework for standardization in telecommunications and information technology, *IEEE Communications Magazine* **39**(4):94–100, 2001.

M. H. Sherif, When standardization is slow? *International Journal of IT Standards & Standardization Research* **1**(1):19–32, 2003a.

M. H. Sherif, Technology substitution and standardization in telecommunications services, in *SIIT 2003 Proceedings 3rd IEEE Conference on Standardization and Innovation in Information Technology,* 22–24 October 2003, Delft, The Netherlands, 2003b, pp. 241–252.

M. H. Sherif, D. Hoeflin, and M. Recchia, Reliability assessment of network elements using black box testing, in *Proceedings of the 7th IEEE Symposium on Computers and Communications* ISCC 2002, 1–4 July 2002, Taormina-Giardini Naxos, Italy, pp. 1015–1200.

M. H. Sherif, D. Hoeflin, and M. Recchia, Risk management for new service introduction in telecommunications networks, in *Proceedings of the 8th IEEE Symposium on Computers and Communications* ISCC 2003, June 30–July 3, 2003, Kemer-Antalya, Turkey, pp. 597–601.

N. Sirén, *Risk Management in Telecommunications,* MBA dissertation, Buckinghamshire Chilterns University College, Chalfont St. Giles, United Kingdom, May 2003.

C. Sorti, *The Art of Coming Home,* Intercultural Press, Yarmouth, ME, 1997.

H. Sparrow, EVM = EVM: Earned Value Management results in Early Visibility and Management opportunities, in *Proceedings of the Project Management Institute Annual Seminars and Symposium,* September 7–16, 2000, Houston, TX.

A. G. Stephenson, CEO's corner. It's about people, *Engineering Management Journal* **14**(1):3–5, 2003.

J. M. Sussman, ITS: What we know now that we wish we knew then—A retrospective on the ITS 1992 strategic plan, Working Papers Series, ESD-WP-2003-09, Massachusetts Institute of Technology, September 2003.

P. Taylor, Costs rise in Cingular, AT&T Wireless deal, *Financial Times,* Nov. 7, 2004, p. 9.

Telcordia Technologies, GR-929-CORE *Telcordia Technologies Generic Requirement—Reliability and quality measurements for telecommunications systems (RQMS-Wireline),* Issue 8, Dec. 2002, Red Bank, NJ.

H. J. Thamhain, Team management: Working effectively, in *Project Management for Business Professionals: A Comprehensive Guide,* J. Knutson, ed., John Wiley & Sons, New York, 2001, pp. 550–572.

H. J. Thamhain, Team building, Chapter 10 in D. Z. Milosevic, *Project Management Toolbox: Tools and Techniques for the Practicing Project Manager,* John Wiley & Sons, Hoboken, NJ, 2003.

J. A. Thoren, Jr., Application of earned value concepts to non-government contracts, in *Proceedings of the Project Management Institute Annual Seminars and Symposium,* September 7–16, 2000, Houston, TX.

J. Thorp, *The Information Paradox: Realizing The Business Benefits of Information Technology,* McGraw-Hill, Toronto, 1998.

J. E. Tingstad, *How to Manage the R&D Staff: A Looking-Glass World,* American Management Association, New York, 1991.

A. Toffler, *Future Shock,* Bantam Books, New York, 1971.

F. Trompenaars and C. Hampdend-Turner, *Riding the Waves of Culture: Understanding Cultural Diversity in Global Business,* 2nd edition, McGraw-Hill, New York, 1998.

S. Tzu, *The Art of War,* translated by Samuel B. Griffith, Oxford University Press, London, 1963.

United Nations—Economic Commission for Europe, *The Telecommunication Industry—Growth and Structural Change,* New York, 1987.

G. Vallet, *Techniques de Planification de Projets,* Dunod, Paris, 2003.

P. Verveer, Telecommunications and the Olympic Games, *IEEE Communications Magazine* **39**(7):69–70, 2001.

Wall Street Journal, Companies seek two separate phone systems—just in case, December 20, 2001.

E. Ward, *World-Class Service Development,* Artech House, Boston, 1998.

J. L. Ward, *Project Management Terms: A Working Glossary,* ESI, Arlington, VA, 2000.

J. West, Institutional constraints in the initial deployment of cellular service on three continents, in *Information Technology Standards and Standardization: A Global Perspective,* IDEA Group Publishing, Hershey, PA, 2000, pp. 198–221.

R. E. Westney, Risk management: Maximizing the probability of success, in *Project Management for Business Professionals: A Comprehensive Guide,* J. Knutson, ed., John Wiley & Sons, New York, 2001, pp. 128–150.

J. J. Wheatley, *World Telecommunications Economics,* IEE, Herts, United Kingdom, 1999.

R. M. Wideman, Project teamwork, personality profiles and the population at large: Do we have enough of the right kind of people? in *Proceedings of the 29th Annual Project Management Institute 1998 Seminars and Symposium,* 1998, pp. 1063–1070.

R. M. Wideman and A. J. Shenhar, Professional and personal develoment management: A practical approach to education and training, in *Project Management for Business Professionals: A Comprehensive Guide,* J. Knutson, ed., John Wiley & Sons, New York, 2001, pp. 353–383.

F. R. Wilcox, Net 1000 Retrospective, An informal summary, Report to the president of AT&T Bell Labs, December 21, 1987.

G. Winch, N. Clifton and C. Millar, Organisation and management in an Anglo-French consortium: the case of Transmanche-Link, *Journal of Management Studies* **37**(5):663–685, 2000.

M. Witzel, A need for the right stuff, *Financial Times,* August 11, 2004, p. 7.

L. A. Wrobel, *Disaster Recovery Pplanning for Telecommunications,* Artech House, Norwood, MA, 1990.

S. Yamada, M. Ohba, and S. Osaki, S-Shaped reliability growth modeling for software error detection, *IEEE Transactions on Reliability* **32**(5):475–478; 484, 1983.

INDEX